Satisficing Games and Decision Making

In our day to day lives we constantly make decisions that are simply "good enough" rather than optimal – a type of decision for which Professor Wynn Stirling has adopted the word "satisficing." Most computer-based decision-making algorithms, on the other hand, doggedly seek only the optimal solution based on rigid criteria, and reject any others. In this book, Professor Stirling outlines an alternative approach, using novel algorithms and techniques which can be used to find satisficing solutions. Building on traditional decision and game theory, these techniques allow decision-making systems to cope with more subtle situations where self and group interest conflict, perfect solutions can't be found and human issues need to be taken into account – in short, more closely modeling the way humans make decisions. The book will therefore be of great interest to engineers, computer scientists, and mathematicians working on artificial intelligence and expert systems.

Wynn C. Stirling is a Professor of Electrical and Computer Engineering at Brigham Young University, where he teaches stochastic processes, control theory, and signal processing. His research interests include decision theory, multi-agent control theory, detection and estimation theory, information theory, and stochastic processes.

T0185909

Satisficing Games and Decision Making

With applications to engineering and computer science

Wynn C. Stirling

Brigham Young University

CAMBRIDGE
UNIVERSITY PRESS

CAMBRIDGE UNIVERSITY PRESS
Cambridge, New York, Melbourne, Madrid, Cape Town, Singapore, São Paulo

Cambridge University Press
The Edinburgh Building, Cambridge CB2 8RU, UK

Published in the United States of America by Cambridge University Press, New York

www.cambridge.org
Information on this title: www.cambridge.org/9780521817240

First published 2003
This digitally printed version 2007

A catalogue record for this publication is available from the British Library

Library of Congress Cataloguing in Publication data
Stirling, Wynn C.
Satisficing games and decision making : with applications to engineering and computer
science / Wynn C. Stirling.
    p.   cm.
Includes bibliographical references and index.
ISBN 0 521 81724 2
1. Decision making.   2. Artificial intelligence.   3. Human-computer interaction.   I. Title.
T57.95 .S73   2003
003'.56 – dc21   2002073606

ISBN 978-0-521-81724-0 hardback
ISBN 978-0-521-03891-1 paperback

For Patti,
whose abundance mentality
provides much more than mere encouragement

Contents

9   Meliority   205

Appendices

Figures

Tables

Alles Gescheite ist schon gedacht worden; man muss nur versuchen, es noch einmal zu denken.

Everything imaginative has been thought before; one must only attempt to think it again.

Johann Wolfgang von Goethe
Maximen und Reflexionen (1829)

Preface

It is the profession of philosophers to question platitudes that others accept without thinking twice. A dangerous profession, since philosophers are more easily discredited than platitudes, but a useful one. For when a good philosopher challenges a platitude, it usually turns out that the platitude was essentially right; but the philosopher has noticed trouble that one who did not think twice could not have met. In the end the challenge is answered and the platitude survives, more often than not. But the philosopher has done the adherents of the platitude a service: he has made them think twice.

David K. Lewis, *Convention* (Harvard University Press, 1969)

It is a platitude that decisions should be optimal; that is, that decision makers should make the best choice possible, given the available knowledge. But we cannot rationally choose an option, even if we do not know of anything better, unless we know that it is good enough. Satisficing, or being "good enough," is the fundamental desideratum of rational decision makers – being optimal is a bonus.

Can a notion of being "good enough" be defined that is distinct from being best? If so, is it possible to formulate the concepts of being good enough for the group and good enough for the individuals that do not lead to the problems that exist with the notions of group optimality and individual optimality? This book explores these questions. It is an invitation to consider a new approach to decision theory and mathematical games. Its purpose is to supplement, rather than supplant, existing approaches. To establish a seat at the table of decision-making ideas, however, it challenges a widely accepted premise of conventional decision theory; namely, that a rational decision maker must always seek to do, and only to do, what is best for itself.

Optimization is the mathematical instantiation of individual rationality, which is the doctrine of exclusive self-interest. In group decision-making settings, however, it is generally not possible to optimize simultaneously for all individuals. The prevailing interpretation of individual rationality in group settings is for the participants to seek an equilibrium solution, where no single participant can improve its level of satisfaction by making a unilateral change. The obvious desirability of optimization and equilibration, coupled with a convenient mathematical formalization via calculus, makes this view of rational choice a favorite of many disciplines. It has served many decision-making communities well for many years and will continue to do so. But there is some disquiet on the horizon. There is a significant movement in engineering and computer science toward "intelligent decision-making," which is an attempt to build machines that mimic, either biologically or cognitively, the processes of human decision making, with the goal of synthesizing artificial entities that possess some of the decision-making power of human beings. It is well documented, however, that humans are poor optimizers, not only because they often cannot be, because of such things as computational and memory limitations, but because they may not care to be, because of their desire to accommodate the interests of others as well as themselves, or simply because they are content with adequate performance. If we are to synthesize autonomous decision-making agents that mimic human behavior, they in all likelihood will be based on principles that are less restrictive than exclusive self-interest.

Cooperation is a much more sophisticated concept than competition. Competition is the natural result of individual rationality, but individual rationality is the Occam's razor of interpersonal interaction, and relies only upon the minimal assumption that an individual will put its own interests above everything and everyone else. True cooperation, on the other hand, requires decision makers to expand their spheres of interest and give deference to others, even at their own expense. True cooperation is very difficult to engender with individual rationality.

Relaxing the demand for strict optimality as an ideal opens the way for consideration of a different principle to govern behavior. A crucial aspect of any decision problem is the notion of balance, such that a decision maker is able to accommodate the various relationships that exist between it and its environment, including other participants. An artificial society that coordinates with human beings must be ecologically balanced to the human component if humans are to be motivated to use and trust it. Furthermore, effective non-autocratic societies must be socially balanced between the interests of the group and the interests of the individuals who constitute the group. Unfortunately, exclusive self-interest does not naturally foster these notions of balance, since each participant is committed to tipping the scale in its own favor, regardless of the effect on others. Even in non-competitive settings this can easily lead to selfish, exploitive, and even avaricious behavior, when cooperative, unselfish, and even altruistic behavior would be more appropriate. This type of behavior can be antisocial and counterproductive, especially if the other participants are not motivated by the same narrow ideal. Conflict cannot be avoided in general, but conflict can just as easily lead to collaboration as to competition.

One cannot have degrees or grades of optimality; either an option is optimal or it is not. But common sense tells us that not all non-optimal options are equal. One of the most influential proponents of other-than-optimal approaches to decision making is Herbert Simon, who appropriated the term "satisficing" to describe an attitude of taking action that satisfies the minimum requirements necessary to achieve a particular goal. Since these standards are chosen arbitrarily, Simon's approach has often been criticized as *ad hoc*. There have been several attempts in the literature to rework his original notion of satisficing into a form of constrained optimization, but such attempts are not true to Simon's original intent. In Chapter 1 Simon's notion of satisficing is retooled by introducing a notion of "good enough" in terms of intrinsic, rather than extrinsic, criteria, and couching this procedure in a new notion of rationality that is termed *intrinsic rationality*.

For a decision maker truly to optimize, it must possess all of the relevant facts. In other words, the localization of interest (individual rationality) seems to require the globalization of preferences, and when a total ordering is not available, optimization is frustrated. Intrinsic rationality, however, does not require a total ordering, since it does not require the global rank-ordering of preferences. In Chapter 2 I argue that forming *conditional local preference orderings* is a natural way to synthesize emergent total orderings for the group as well as for the individual. In other words, the localization of preferences can lead to the globalization of interest.

The desire to consider alternatives to traditional notions of decision-making has also been manifest in the philosophical domain. In particular, Isaac Levi has challenged traditional epistemology. Instead of focusing attention on justifying existing knowledge, he concentrates on how to improve knowledge. He questions the traditional goal of seeking the truth and nothing but the truth and argues that a more modest and achievable goal is that of seeking new information while avoiding error. He offers, in clean-cut mathematical language, a framework for making such evaluations. The result is Levi's *epistemic utility theory.*

Epistemology involves the classification of propositions on the basis of knowledge and belief regarding their content, and praxeology involves the classification of options on the basis of their effectiveness. Whereas epistemology deals with the issue of what to believe, praxeology deals with the issue of how to act. The praxeic analog to the conventional epistemic notion of seeking the truth and nothing but the truth is that of taking the best and nothing but the action. The praxeic analog to Levi's more modest epistemic goal of acquiring new information while avoiding error is that of conserving resources while avoiding failure. Chapter 3 describes a transmigration of Levi's original philosophical ideas into the realm of practical engineering. To distinguish between the goals of deciding what to believe and how to act, this reoriented theory is termed *praxeic utility theory.*

Praxeic utility theory provides a definition for satisficing decisions that is consistent with intrinsic rationality. Chapter 4 discusses some of the properties of satisficing decisions and introduces the notion of *satisficing equilibria* as a refinement of the fundamental satisficing concept. It also establishes some fundamental consistency relationships.

Chapter 5 addresses two kinds of uncertainty. The first is the usual notion of *epistemic uncertainty* caused by the lack of knowledge and is usually characterized with probability theory. The second kind of uncertainty is termed *praxeic uncertainty* and deals with the equivocation and sensitivity that a decision maker may experience as a result of simply being thrust into a decision-making environment. Praxeic uncertainty deals with the innate ability of the decision maker.

One of the main benefits of satisficing *à la* praxeic utility theory is that it admits a natural extension to a community of decision makers. Chapter 6 presents a theory of multi-agent decision making that is very different from conventional von Neumann–Morgenstern game theory, which focuses on maximizing individual expectations conditioned on the actions of other players. This new theory, termed *satisficing game theory*, permits the direct consideration of group interests as well as individual interests and mitigates the attitude of competition that is so prevalent in conventional game theory.

Negotiation is one of the most difficult and sophisticated aspects of N-person von Neumann–Morgenstern game theory. One of the reasons for this difficulty is that the principle of individual rationality does not permit a decision maker to enter into compromise agreements that would permit any form of self-sacrifice, no matter how slight for the person, or how beneficial it may be for others. Chapter 7 shows how satisficing does permit such behavior and possesses a mechanism to control the degree of self-sacrifice that a decision maker would permit when attempting to achieve a compromise.

Multi-agent decision-making is inherently complex. Furthermore, praxeic utility theory leads to more complexity than does standard von Neumann–Morgenstern game theory, but it is not more complex than it needs to be to characterize all multi-agent preferences. Chapter 8 demonstrates this increased complexity by recasting some well-known games as satisficing games and discusses modeling assumptions that can mitigate complexity.

Chapter 9 reviews some of the distinctions between satisficing and optimization, discusses the ramifications of choosing the rationality criterion, and extends an invitation to examine some significant problems from the point of view espoused herein.

Having briefly described what this book is about, it is important also to stress what it is *not* about. It is not about a social contract (i.e., the commonly understood coordinating regularities by which a society operates) to characterize *human* behavior. Lest I be accused of heresy or, worse, naiveté by social scientists, I wish to confine my application to the synthesis of *artificial* decision-making societies. I employ the arguments of philosophers and social scientists to buttress my claim that any "social contract" for artificial systems should not be confined to the narrow precepts of individual rationality, but I do not claim that the notion of rationality I advance is *the* explanation for human social behavior. I do believe, however, that it is compatible with human behavior and should be considered as a component of any man–machine "social contract" that may eventually emerge as decision-making machines become more sophisticated and the interdependence of humans and machines increases.

This book had its beginnings several years ago. While a graduate student I happened to overhear a remark from a respected senior faculty member, who lamented, as nearly as I can recall, that "virtually every PhD dissertation in electrical engineering is an application of X dot equals zero [$\dot{X} = 0$]." He was referring to an elementary theorem from calculus that functions achieve their maxima and minima at points where the derivative vanishes. Sophisticated versions of this basic idea are the mainstays of optimization-based methods. Before hearing that remark, it had never occurred to me to question the standard practice of optimization. I had taken for granted that, without at least some notion of optimality, decision-making would be nothing more than an exercise in adhocism. I was nevertheless somewhat deflated to think that my own dissertation, though garnished with some sophisticated mathematical accoutrements, was really nothing more, at the end of the day, than yet another application of $\dot{X} = 0$. Although this realization did not change my research focus at the time, it did eventually prompt me to evaluate the foundational assumptions of decision theory.

I am not a critic of optimization, but I am a critic of indiscriminately prescribing it for all situations. Principles should not be adopted simply out of habit or convenience. If one of the goals of philosophy is to increase contact with reality, then engineers, who seek not only to appreciate reality but also to create it, should occasionally question the philosophical underpinnings of their discipline. This book expresses the hope that the cultures of philosophy and engineering can be better integrated. Good designs should be based on good philosophy and good philosophy should lead to good designs. The philosophy of "good enough" deserves a seat at the table alongside the philosophy of "nothing but the best." Neither is appropriate for all situations. Both have their limitations and their natural applications. Satisficing, as a precisely defined mathematical concept, is another tool for the decision maker's toolbox.

This book was engendered through many fruitful associations. Former students Darryl Morrell and Mike Goodrich have inspired numerous animated and stimulating discussions as we hammered out many of the concepts that have found their way into this book. Fellow engineers and collaborators Rick Frost, Todd Moon, and Randy Beard have been unfailing sources of enlightenment and encouragement. Also, Dennis Packard and Hal Miller of the philosophy and psychology departments, respectively, at BYU, have helped me to appreciate the advantages of collaboration between engineering, the humanities, and the social and behavorial sciences. In particular, I owe a special debt of gratitude to Hal, who carefully critiqued the manuscript and made many valuable suggestions.

1 Rationality

> Rationality, according to some, is an excess of reasonableness. We should be rational enough to confront the problems of life, but there is no need to go whole hog. Indeed, doing so is something of a vice.
>
> Isaac Levi, *The Covenant of Reason* (Cambridge University Press, 1997)

The disciplines of science and engineering are complementary. Science comes from the Latin root *scientia*, or knowledge, and engineering comes from the Latin root *ingenerare*, which means to beget. While any one individual may fulfill multiple roles, a scientist *qua* seeker of knowledge is concerned with the analysis of observed natural phenomena, and an engineer *qua* creator of new entities is concerned with the synthesis of artificial phenomena. Scientists seek to develop models that explain past behavior and predict future behavior of the natural entities they observe. Engineers seek to develop models that characterize desired behavior for the artificial entities they construct. Science addresses the question of how things are; engineering addresses the question of how things might be.

Although of ancient origin, science as an organized academic discipline has a history spanning a few centuries. Engineering is also of ancient origin, but as an organized academic discipline the span of its history is more appropriately measured by a few decades. Science has refined its methods over the years to the point of great sophistication. It is not surprising that engineering has, to a large extent, appropriated and adapted for synthesis many of the principles and techniques originally developed to aid scientific analysis.

One concept that has guided the development of scientific theories is the "principle of least action," advanced by Maupertuis[1] as a means of systematizing Newtonian mechanics. This principle expresses the intuitively pleasing notion that nature acts in a way that gives the greatest effect with the least effort. It was championed by Euler, who said: "Since the fabric of the world is the most perfect and was established by the wisest Creator, nothing happens in this world in which some reason of maximum or minimum

---

[1] Beeson (1992) cites Maupertuis (1740) as Maupertuis' first steps toward the development of this principle.

would not come to light" (quoted in Polya (1954)).[2] This principle has been adopted by engineers with a fruitful vengeance. In particular, Wiener (1949) inaugurated a new era of estimation theory with his work on optimal filtering, and von Neumann and Morgenstern (1944) introduced a new structure for optimal multi-agent interactivity with their seminal work on game theory. Indeed, we might paraphrase Euler by saying: "Nothing should be designed or built in this world in which some reason of maximum or minimum would not come to light." To obtain credibility, it is almost mandatory that a design should display some instance of optimization, even if only approximately. Otherwise, it is likely to be dismissed as *ad hoc*.

However, analysis and synthesis are inverses. One seeks to take things apart, the other to put things together. One seeks to simplify, the other to complicate. As the demands for complexity of artificial phenomena increase, it is perhaps inevitable that principles and methods of synthesis will arise that are not attributable to an analysis heritage – in particular, to the principle of least action. This book proposes such a method. It is motivated by the desire to develop an approach to the synthesis of artificial multi-agent decision-making systems that is able to accommodate, in a seamless way, the interests of both individuals and groups.

Perhaps the most important (and most difficult) social attribute to imitate is that of coordinated behavior, whereby the members of a group of autonomous distributed machines coordinate their actions to accomplish tasks that pursue the goals of both the group and each of its members. It is important to appreciate that such coordination usually cannot be done without conflict, but conflict need not degenerate to competition, which can be destructive. Competition, however, is often a byproduct of optimization, whereby each participant in a multi-agent endeavor seeks to achieve the best outcome for itself, regardless of the consequences to other participants or to the community.

Relaxing the demand for optimization as an ideal may open avenues for collaboration and compromise when conflict arises by giving joint consideration to the interests of the group and the individuals that compose the group, provided they are willing to accept behavior that is "good enough." This relaxation, however, must not lead to reliance upon *ad hoc* rules of behavior, and it should not categorically exclude optimal behavior. To be useful for synthesis, an operational definition of what it means to be good enough must be provided, both conceptually and mathematically. The intent of this book is two-fold: (a) to offer a criterion for the synthesis of artificial decision-making systems that is designed, from its inception, to model both collective and individual interests; and (b) to provide a mathematical structure within which to develop and apply this criterion. Together, criterion and structure may provide the basis for an alternative view of the design and synthesis of artificial autonomous systems.

---

[2] Euler's argument actually begs the question by using superlatives (most perfect, wisest) to justify other superlatives (maximum, minimum).

1.1 Games machines play

Much research is being devoted to the design and implementation of artificial social systems. The envisioned applications of this technology include automated air-traffic control, automated highway control, automated shop floor management, computer network control, and so forth. In an environment of rapidly increasing computer power and greatly increased scientific knowledge of human cognition, it is inevitable that serious consideration will be given to designing artificial systems that function analogously to humans. Many researchers in this field concentrate on four major metaphors: (a) brain-like models (neural networks), (b) natural language models (fuzzy logic), (c) biological evolutionary models (genetic algorithms), and (d) cognition models (rule-based systems). The assumption is that by designing according to these metaphors, machines can be made at least to imitate, if not replicate, human behavior. Such systems are often claimed to be intelligent.

The word "intelligent" has been appropriated by many different groups and may mean anything from nonmetaphorical cognition (for example, strong AI) to advertising hype (for example, intelligent lawn mowers). Some of the definitions in use are quite complex, some are circular, and some are self-serving. But when all else fails, we may appeal to etymology, which owns the deed to the word; everyone else can only claim squatters rights. *Intelligent* comes from the Latin roots *inter* (between) + *legĕre* (to choose). Thus, it seems that an indispensable characteristic of intelligence in man or machine is an ability to choose between alternatives.

Classifying "intelligent" systems in terms of anthropomorphic metaphors categorizes mainly their syntactical, rather than their semantic, attributes. Such classifications deal primarily with the way knowledge is represented, rather than with the way decisions are made. Whether knowledge is represented by neural connection weights, fuzzy set-membership functions, genes, production rules, or differential equations, is a choice that must be made according to the context of the problem and the preferences of the system designer. The way knowledge is represented, however, does not dictate the rational basis for the way choices are made, and therefore has little to do with that indispensable attribute of intelligence.

A possible question, when designing a machine, is the issue of just where the actual choosing mechanism lies – with the designer, who must supply the machine with all of rules it is to follow, or with the machine itself, so that it possesses a degree of true autonomy (self-governance). This book does not address that question. Instead, it focuses primarily on the issue of *how* decisions might be made, rather than *who* ultimately bears the responsibility for making them. Its concern is with the issue of how to design artificial systems whose decision-making mechanisms are understandable to and viewed as reasonable by the people who interface with such systems. This concern leads directly to a study of rationality.

This book investigates rationality models that may be used by men or machines. A rational decision is one that conforms either to a set of general principles that govern preferences or to a set of rules that govern behavior. These principles or rules are then applied in a logical way to the situation of concern, resulting in actions which generate consequences that are deemed to be acceptable to the decision maker. No single notion of what is acceptable is sufficient for all situations, however, so there must be multiple concepts of rationality. This chapter first reviews some of the commonly accepted notions of rationality and describes some of the issues that arise with their implementation. This review is followed by a presentation of an alternative notion of rationality and arguments for its appropriateness and utility. This alternative is not presented, however, as a panacea for all situations. Rather, it is presented as a new formalism that has a place alongside other established notions of rationality. In particular, this approach to rational decision-making is applicable to multi-agent decision problems where cooperation is essential and competition may be destructive.

1.2   Conventional notions

The study of human decision making is the traditional bailiwick of philosophy, economics, and political science, and much of the discussion of this topic concentrates on defining what it means to have a degree of conviction sufficient to impel one to take action. Central to this traditional perspective is the concept of preference ordering.

Definition 1.1
Let the symbols "$\succeq$" and "$\cong$" denote binary ordering relationships meaning "is at least as good as" and "is equivalent to," respectively. A **total ordering** of a collection of options $U = \{u_1, \ldots, u_n\}$, $n \geq 3$, occurs if the following properties are satisfied:

Reflexivity: $\forall u_i \in U : u_i \succeq u_i$

Antisymmetry: $\forall u_i, u_j \in U : u_i \succeq u_j \ \& \ u_j \succeq u_i \Rightarrow u_i \cong u_j$

Transitivity: $\forall u_i, u_j, u_k \in U : u_i \succeq u_j, \ u_j \succeq u_k \Rightarrow u_i \succeq u_k$

Linearity: $\forall u_i, u_j \in U : u_i \succeq u_j \text{ or } u_j \succeq u_i$

If the linearity property does not hold, the set U is said to be **partially ordered**.   □

Reflexivity means that every option is at least as good as itself, antisymmetry means that if u_i is at least as good as u_j and u_j is at least as good as u_i, then they are equivalent, transitivity means that if u_i is as least as good as u_j and u_j is at least as good as u_k, then u_i is at least as good as u_k, and linearity means that for every u_i and u_j pair, either u_i is at least as good as u_j or u_j is at least as good as u_i (or both).

1.2.1   Substantive rationality

Once in possession of a preference ordering, a rational decision maker must employ general principles that govern the way the orderings are to be used to formulate decision rules. No single notion of what is acceptable is appropriate for all situations, but perhaps the most well-known principle is the classical economics hypothesis of Bergson and Samuelson, which asserts that individual interests are fundamental; that is, that social welfare is a function of individual welfare (Bergson, 1938; Samuelson, 1948). This hypothesis leads to the doctrine of **rational choice**, which is that "each of the individual decision makers behaves as if he or she were solving a constrained maximization problem" (Hogarth and Reder, 1986b, p. 3). This paradigm is the basis of much of conventional decision theory that is used in economics, the social and behavioral sciences, and engineering. It is based upon two fundamental premises.

P-1   *Total ordering*: the decision maker is in possession of a total preference ordering for all of its possible choices under all conditions (in multi-agent settings, this includes knowledge of the total orderings of all other participants).

P-2   *The principle of individual rationality*: a decision maker should make the best possible decision for itself, that is, it should optimize with respect to its own total preference ordering (in multi-agent settings, this ordering may be influenced by the choices available to the other participants).

Definition 1.2

Decision makers who make choices according to the principle of individual rationality according to their own total preference ordering are said to be **substantively rational**.   □

One of the most important accomplishments of classical decision theory is the establishment of conditions under which a total ordering of preferences can be quantified in terms of a mathematical function. It is well known that, given the proper technical properties (e.g., see Ferguson (1967)), there exists a real-valued function that agrees with the total ordering of a set of options.

Definition 1.3

A **utility** ϕ on a set of options U is a real-valued function such that, for all $u_i, u_j \in U$, $u_i \succeq u_j$ if, and only if, $\phi(u_i) \geq \phi(u_j)$.   □

Through utility theory, the qualitative ordering of preferences is made equivalent to the quantitative ordering of the utility function. Since it may not be possible, due to uncertainty, to ensure that any given option obtains, orderings are usually taken with respect to expected utility, that is, utility that has been averaged over all options according to the probability distribution that characterizes them; that is,

$$\pi(u) = E[\phi(u)] = \int_U \phi(u) P_C(du),$$

where $E[\cdot]$ denotes mathematical expectation and P_C is a probability measure characterizing the random behavior associated with the set U. Thus, an equivalent notion for substantive rationality (and the one that is usually used in practice) is to equate it with maximizing expected utility (Simon, 1986).

Not only is substantive rationality the acknowledged standard for calculus/ probability-based knowledge representation and decision making, it is also the *de facto* standard for the alternative approaches based on anthropomorphic metaphors. When designing neural networks, algorithms are designed to calculate the *optimum* weights, fuzzy sets are defuzzified to a crisp set by choosing the element of the fuzzy set with the *highest* degree of set membership, genetic algorithms are designed under the principle of survival of the *fittest*, and rule-based systems are designed according to the principle that a decision maker will operate in its own *best* interest according to what it knows.

There is a big difference in perspective between the activity of analyzing the way rational decision makers make decisions and the activity of synthesizing actual artificial decision makers. It is one thing to postulate an explanatory story that justifies how decision makers might arrive at solution, even though the story is not an explicit part of the generative decision-making model and may be misleading. It is quite another thing to synthesize artificial decision makers that actually live such a story by enacting the decision-making logic that is postulated. Maximizing expectations tells us what we may expect when rational entities function, but it does not give us procedures for their operation. It may be instructive, but it is not constructive.

Nevertheless, substantive rationality serves as a convenient and useful paradigm for the synthesis of artificial decision makers. This paradigm loses some of its appeal, however, when dealing with decision-making societies. The major problem is that maximizing expectations is strictly an individual operation. Group rationality is not a logical consequence of individual rationality, and individual rationality does not easily accommodate group interests (Luce and Raiffa, 1957).

Exclusive self-interest fosters competition and exploitation, and engenders attitudes of distrust and cynicism. An exclusively self-interested decision maker would likely assume that the other decision makers also will act in selfish ways. Such a decision maker might therefore impute self-interested behavior to others that would be damaging to itself, and might respond defensively. While this may be appropriate in the presence of serious conflict, many decision scenarios involve situations where coordinative activity, even if it leads to increased vulnerability, may greatly enhance performance. Especially when designing artificial decision-making communities, individual rationality may not be an adequate principle with which to characterize desirable behavior in a group.

The need to define adequate frameworks in which to synthesize rational decision-making entities in both individual and social settings has led researchers to challenge the traditional models based on individual rationality. One major criticism is the claim that people do not usually conform to the strict doctrine of substantive rationality – they are not utility maximizers (Mansbridge, 1990a; Sober and Wilson, 1998; Bazerman, 1983; Bazerman and Neale, 1992; Rapoport and Orwant, 1962; Slote, 1989). It is not clear, in the presence of uncertainty, that the best possible thing to do is always to choose a decision that optimizes a single performance criterion. Although deliberately opting for less than the best possible leaves one open to charges of capriciousness, indecision, or foolhardiness, the incessant optimizer may be criticized as being restless, insatiable, or intemperate.[3] Just as moderation may tend to stabilize and temper cognitive behavior, deliberately backing away from strict optimality may provide protection against antisocial consequences. Moderation in the short run may turn out to be instrumentally optimal in the long run.

Even in the light of these considerations, substantive rationality retains a strong appeal, especially because it provides a systematic solution methodology, at least for single decision makers. One of the practical benefits of optimization is that by choosing beforehand to adopt the option that maximizes expected utility, the decision maker has completed the actual decision making – all that is left is to solve or search for that option (for this reason, much of what is commonly called decision theory may more accurately be characterized as search theory). This fact can be exploited to implement efficient search procedures, especially with concave and differentiable utility functions, and is a computational benefit of such enormous value that one might be tempted to adopt substantive rationality primarily because it offers a systematic and reliable means of finding a solution.

1.2.2   Procedural rationality

If we were to abandon substantive rationality, what justifiable notion of reasonableness could replace it? If we were to eschew optimization and its attendant computational mechanisms, how would solutions be systematically identified and computed? These are significant questions, and there is no single good answer to them. There is, however, a notion of rationality that has evolved more or less in parallel with the notion of substantive rationality and that is relevant to psychology and computer science.

Definition 1.4
Decision makers who make choices by following specific rules or procedures are said to be **procedurally rational** (Simon, 1986).   □

---

[3] As Epicurus put it: "Nothing is enough for the man to whom enough is too little."

For an operational definition of procedural rationality, we turn to Simon:

> The judgment that certain behavior is "rational" or "reasonable" can be reached only by viewing the behavior in the context of a set of premises or "givens." These givens include the situation in which the behavior takes place, the goals it is aimed at realizing, and the computational means available for determining how the goals can be attained. (Simon, 1986, p. 26)

Under this notion, a decision maker should concentrate attention on the quality of the *processes* by which choices are made, rather than directly on the quality of the outcome. Whereas, under substantive rationality, attention is focused on *why* decision makers should do things, under procedural rationality attention is focused on *how* decision makers should do things. Substantive rationality tells us where to go, but not how to get there; procedural rationality tells us how to get there, but not where to go. Substantive rationality is viewed in terms of the outcomes it produces; procedural rationality is viewed in terms of the methods it employs.

Procedures are often heuristic. They may involve *ad hoc* notions of desirability, and they may simply be rules of thumb for selective searching. They may incorporate the same principles and information that could be used to form a substantively rational decision, but rather than dictating a specific option, the criteria are used to guide the decision maker by identifying patterns that are consistent with its context, goals, and computational capabilities.[4] A fascinating description of heuristics and their practical application is found in Gigerenzer and Todd (1999). Heuristics are potentially very powerful and can be applied to more complex and less well structured problems than traditional utility maximization approaches. An example of a procedurally rational decision-making approach is a so-called *expert system*, which is typically composed of a number of rules that specify behavior in various local situations. Such systems are at least initially defined by human experts or authorities.

The price for working with heuristics is that solutions cannot in any way be construed as optimal – they are functional at best. In contrast to substantively rational solutions, which enjoy an absolute guarantee of maximum success (assuming that the model is adequate – we should not forget that "experts" defined these models as well), procedurally rational solutions enjoy no such guarantee.

A major difference between substantive rationality and procedural rationality is the capacity for self-criticism, that is, the capacity for the decision maker to evaluate its own performance in terms of coherence and consistency. Self-criticism will be built into substantive rationality if the criteria used to establish optimality can also be used

---

[4] A well-known engineering example of the distinction between substantive rationality and procedural rationality is found in estimation theory. The so-called Wiener filter (Wiener, 1949) is the substantively rational solution that minimizes the mean-square estimation error of a time-invariant linear estimator. However, the performance of the Wiener filter is often approximated by a heuristic, called the LMS (least-mean-square) filter and developed by Widrow (1971). Whereas the Wiener filter is computed independently of the actual observations, the Widrow filter is generated by the observations. The Wiener filter requires that all stochastic processes be stationary and modeled to the second order; the Widrow filter relaxes those constraints. Both solutions are extremely useful in their appropriate settings, but they differ fundamentally.

to define the search procedure.[5] By contrast, procedural rationality does not appear to possess a self-policing capacity. The quality of the solution depends on the abilities of the expert who defined the heuristic, and there may be no independent way to ascribe a performance metric to the solution from the point of view of the heuristic. Of course, it is possible to apply performance criteria to the solution once it has been identified, but such *post factum* criteria do not influence the choice, except possibly in conjunction with a learning mechanism that could modify the heuristics for future application. While it may be too strong to assert categorically that heuristics are incapable of self-criticism, their ability to do so on a single trial is at least an open question.

Substantive rationality and procedural rationality represent two extremes. On the one hand, substantive rationality requires the decision maker to possess a complete understanding of the environment, including knowledge of the total preference orderings of itself and all other agents in the group. Any uncertainty regarding preferences must be expressed in terms of expectations according to known probability distributions. Furthermore, even given complete understanding, the decision maker must have at its disposal sufficient computational power to identify an optimal solution. Substantive rationality is highly structured, rigid, and demanding. On the other hand, procedural rationality involves the use of heuristics whose origins are not always clear and defensible, and it is difficult to predict with assurance how acceptable the outcome will be. Procedural rationality is amorphous, plastic, and somewhat arbitrary.

1.2.3 Bounded rationality

Many researchers have wrestled with the problem of what to do when it is not possible or expedient to obtain a substantively rational solution due to informational or computational limitations. Simon identified this predicament when he introduced the notion of **satisficing**.[6]

Because real-world optimization, with or without computers, is impossible, the real economic actor is in fact a satisficer, a person who accepts "good enough" alternatives, not because less is preferred to more, but because there is no choice. (Simon, 1996, p. 28)

To determine whether an alternative is "good enough," there must be some way to evaluate its quality. Simon's approach is to determine quality according to the criteria used for substantive rationality, and to evaluate quality against a standard (the aspiration level) that is chosen more or less arbitrarily. Essentially, one continues searching for an optimal choice until an option is identified that meets the decision maker's aspiration level, at which point the search may terminate.

---

[5] This will be the case if the optimality existence proof is constructive. A non-constructive example, however, is found in information theory. Shannon capacity is an upper bound on the rate of reliable information transmission, but the proof that an optimal code exists does not provide a coding scheme to achieve capacity.

[6] This term is actually of ancient origin (*circa* 1600) and is a Scottish variant of satisfy.

The term "satisficing," as used by Simon, comprises a blend of the two extremes of substantive and procedural rationality and is a species of what he termed **bounded rationality**. This concept involves the exigencies of practical decision making and takes into consideration the informational and computational constraints that exist in real-world situations.

There are many excellent treatments of bounded rationality (see, e.g., Simon (1982a, 1982b, 1997) and Rubinstein (1998)). Appendix A provides a brief survey of the mainstream of bounded rationality research. This research represents an important advance in the theory of decision making; its importance is likely to increase as the scope of decision-making grows. However, the research has a common theme, namely, that if a decision maker could optimize, it surely should do so. Only the real-world constraints on its capabilities prevent it from achieving the optimum. By necessity, it is forced to compromise, but the notion of optimality remains intact. Bounded rationality is thus an approximation to substantive rationality, and remains as faithful as possible to the fundamental premises of that view.

I also employ the term "satisficing" to mean "good enough." The difference between the way Simon employs the term and the way I use it, however, is that satisficing *à la* Simon is an approximation to being best (and is constrained from achieving this ideal by practical limitations), whereas satisficing as I use it is treats being good enough as the ideal (rather than an approximation).

This book is not about bounded rationality. Rather, I concentrate on evaluating the appropriateness of substantive and procedural rationality paradigms as models for multi-agent decision making, and provide an alternative notion of rationality. The concepts of boundedness may be applied to this alternative notion in ways similar to how they are currently applied to substantive rationality, but I do not develop those issues here.

1.3   Middle ground

Substantive rationality is the formalization of the common sense idea that one should do the best thing possible and results in perhaps the strongest possible notion of what should constitute a reasonable decision – the only admissible option is the one that is superior to all alternatives. Procedural rationality is the formalization of the common sense idea that, if something has worked in the past, it will likely work in the future and results in perhaps the weakest possible notion of what should constitute a reasonable decision – an option is admissible if it is the result of following a procedure that is considered to be reliable. Bounded rationality is a blend of these two extreme views of rational decision making that modifies the premises of substantive rationality because of a lack of sufficient information to justify strict adherence to them.

Instead of merely blending the two extreme views of rational decision making, however, it may be useful to consider a concept of rationality that is not derived from either the doctrine of rational choice or heuristic procedures. Kreps seems to express a desire along these lines when he observes that:

... the real accomplishment will come in finding an interesting middle ground between hyperrational behaviour and too much dependence on *ad hoc* notions of similarity and strategic expectations. When and if such a middle ground is found, then we may have useful theories for dealing with situations in which the rules are somewhat ambiguous. (Kreps, 1990, p. 184)

Is there really a middle ground, or is the lacuna between strict optimality and pure heuristics bridgeable only by forming an *ad hoc* hybrid of these extremes? If a non-illusory middle ground does exist, it is evident that few have staked formal claims to any of it. The literature involving substantive rationality (bounded or unbounded), particularly in the disciplines of decision theory, game theory, optimal control theory, and operations research, is overwhelmingly vast, reflecting many decades of serious research and development. Likewise, procedural rationality, in the form of heuristics, rule-based decision systems, and various *ad hoc* techniques, is well-represented in the computer science, social science, and engineering literatures. Also, the literature on bounded rationality as a modification or blend of these two extremes is growing at a rapid pace. Work involving rationality paradigms that depart from these classical views, however, is not in evidence.

One of the goals of this book is to search not only for middle ground but for new turf upon which to build. In doing so, let us first examine a "road map" that may guide us to fruitful terrain. The map consists of desirable attributes of the notion of rationality we seek.

A-1 *Adequacy*: satisficing, or being "good enough," is the fundamental desideratum of rational decision makers. We cannot rationally choose an option, even when we do not know of anything better, unless we know that it is good enough. Insisting on the best and nothing but the best, however, can be an unachievable luxury.

A-2 *Sociality*: rationality must be defined for groups as well as for individuals in a consistent and coherent way, such that both group and individual preferences are accommodated. Group rationality should not be defined in terms of individual rationality nor vice versa.

These attributes represent a general relaxing of substantive rationality. Liberation from maximization may open the door to accommodating group as well as individual interests, while still maintaining the integrity supplied by adherence to principles. The attributes also bring rigor to procedural rationality, since they move away from purely *ad hoc* methods and insist on the capacity for self-criticism.

1.3.1   Adequacy

Adequacy is a harder concept to deal with than optimality. Achieving the summit of a mountain is a simple concept that does not depend upon the valley below. By contrast, getting high enough to see across the valley depends upon the valley as well as the mountain. Optimality can be considered objective and is abstracted from context, but adequacy is subjective, that is, it is context dependent. Abstractification is powerful. It transforms a messy real-world situation into a clean mathematical expression that permits the power of calculus and probability theory to be focused on finding a solution. The advantages of abstractification are enormous and not lightly eschewed, and their appeal has fundamentally changed the way decision-making is performed in many contexts. But Zadeh, the father of fuzzy logic, suggests that always insisting on optimality is shooting beyond the mark, and that a softer notion of what is reasonable must be considered.

> Not too long ago we were content with designing systems which merely met given specifications... Today, we tend, perhaps, to make a fetish of optimality. If a system is not the "best" in one sense or another, we do not feel satisfied. Indeed, we are apt to place too much confidence in a system that is, in effect, optimal by definition...
>
> At present, no completely satisfactory rule for selecting decision functions is available, and it is not very likely that one will be found in the foreseeable future. Perhaps all that we can reasonably expect is a rule which, in a somewhat equivocal manner, would delimit a set of "good" designs for a system. (Zadeh, 1958)

A clear operational definition for what it means to be satisficing, or good enough, must be a central component of the notion of rationality that we are seeking. Zadeh reminds us that no such notion is likely to be a panacea, and any definition we offer is subject to criticism and must be used with discretion. Indeed, decision making is inherently equivocal, as uncertainty can never be completely eliminated.

To make progress in our search for what it means to be good enough, we must be willing to relax the demand for strict optimality. We should not, however, abandon the criteria that are used to define optimality, but only the demand to focus attention exclusively on the optimal solution. We certainly should not contradict the notion of optimality by preferring options that are poor according to the optimality criteria over those that comply with the criteria. The goal is to give place to a softer notion of rationality that accommodates, in a formal way, the notion of being good enough.

To maintain the criteria of optimality but yet not insist on optimality may seem paradoxical. If we know what is best, what possible reason could there be for not choosing it? At least a partial answer is that optimization is an ideal that serves to guide our search for an acceptable choice, but not necessarily to dictate what the final choice is. For example, when I drive to work my criterion is to get there in a timely manner, but I do not need to take the quickest route to satisfy the criterion. Strict optimality does not let me consider any but the very best route.

It is not irrational, in the view of some philosophers, for people not to optimize. Slote, for example, argues that it is reasonable not only to settle for something that is less than the best, but that such a situation may actually be preferred by a rational decision maker. That is, one may willfully and rationally eschew taking the action that maximizes utility.

> Defenders of satisficing claim that it sometimes makes sense not to pursue one's own greatest good or desire-fulfillment, but I think it can also be shown that it sometimes makes sense deliberately to reject what is *better* for oneself in favor of what is *good and sufficient* for one's purposes. Those who choose in this way demonstrate a modesty of desire, a kind of moderation, that seems intuitively understandable, and it is important to gain a better understanding of such moderation if we wish to become clear, or clearer, about common-sense, intuitive rationality. (Slote, 1989, pp. 1–2; emphasis in original)

The gist of Slote's argument is that common sense rationality differs from optimizing views of rationality in a way analogous to the difference between common sense morality and utilitarian views of deontology. According to this latter view, what one is morally permitted to do, one is morally required to do. Similarly, substantive rationality requires one to optimize if one is able to do so. Slote argues that, just as utilitarian deontology prohibits decision makers from acting supererogatorily, that is, of doing more than is required or expected, optimizing views of rationality prohibit one from achieving less than one is capable of achieving. But common sense morality permits supererogation, and common sense rationality permits moderation.

Although Slote criticizes optimization as a model for behavior, he does not provide an explicit criterion for characterizing acceptable other-than-optimal activity. While an explicit criterion may not be necessary in the human context, when designing artificial agents, the designer must provide them with some operational mechanism to govern their decision-making if they are to function in a coherent way. Perhaps the weakest notion of rationality that would permit such activity is an operational notion of being "good enough."

One way to establish what it means to be good enough is to specify minimum requirements and accept any option that meets them. This is the approach taken by Simon. He advocates the construction of "aspiration levels" and to halt searching when they are met (Simon, 1955). Although aspiration levels at least superficially establish minimum requirements, this approach relies primarily upon experience-derived expectations. If the aspiration is too low, something better may needlessly be sacrificed, and if it is too high, there may be no solution. It is difficult to establish an adequate practically attainable aspiration level without first exploring the limits of what is possible, that is, without first identifying optimal solutions – the very activity that satisficing is intended to circumvent.[7] Furthermore, such an approach is susceptible to the charge that defining "good enough" in terms of minimum requirements begs the question, because the only way seemingly to define minimum requirements is that they are good enough.

---

[7] The decision maker may, however, be able to adjust his or her aspirations according to experience (see Cyert and March (1992)), in which case it may be possible to adopt aspiration levels that are near-optimal. Even so, however, there may be no way to determine how far one is away from the optimal solution without searching directly for it.

For single-agent low-dimensional problems, specifying the aspirations may be non-controversial. But, with multi-agent systems, interdependence between decision makers can be complex, and aspiration levels can be conditional (what is satisfactory for me may depend upon what is satisfactory for you).

Satisficing via aspiration levels involves making a tradeoff between the cost of continuing to search for a better solution than one currently has and the adequacy of the solution already in hand. That is, for any option under consideration, the decision maker makes a choice between accepting the option and stopping the search or rejecting the option and continuing the search. Making decisions in this way is actually quite similar to the way decisions are made under substantive rationality; it is only the stopping rule that is different. Both approaches rank-order the options and stop when one is found with acceptably high rank. With optimality, the ranking is relative to other options, and searching stops when the highest-ranking option is found. With aspiration levels, the ranking is done with respect to an externally supplied standard, and searching stops when an option is found whose ranking exceeds this threshold.

What aspiration levels and optimization have in common is that the comparison operation is *extrinsic*, that is, the ranking of a given option is made with respect to attributes that are not necessarily part of the option. In the case of optimization, comparisons are made relative to other options. In the case of aspiration levels, comparisons are made relative to an externally supplied standard. Under both paradigms, an option is selected or rejected on the basis of how it compares to things external to itself. Also, both rank-order comparisons and fixed-standard comparisons are global, in that each option is categorized in the option space relative to all other options.

Total ordering, however, is not the only way to make comparisons, nor is it the most fundamental way. A more primitive approach is to form dichotomies, that is, to define two distinct (and perhaps conflicting) sets of attributes for each option and either to select or reject the option on the basis of comparing these attributes. Such dichotomous comparisons are *intrinsic*, since they do not necessarily reference anything not directly relating to the option.

Whereas extrinsic decisions are of the form: either select Hamburger A or select Hamburger B (presumably on the basis of appearance and cost), intrinsic decisions are of the form: either select Hamburger A or reject Hamburger A, with a similar decision required for Hamburger B. The difference is that, under the extrinsic model, one would combine appearance and cost into a single utility that could be rank-ordered, but under the intrinsic model, one forms the binary evaluation of appearance versus cost. If only one of the hamburgers passes muster, the problem is resolved. If you conclude that neither hamburger's appearance is worthy of the cost, you are justified in rejecting them both. If you think both are worthy but you must choose only one, then you either may appeal to a more sophisticated (e.g., extrinsic) decision paradigm, or you may include additional criteria and try again, or you may make a random choice between the options. Suppose that Hamburger A costs more than Hamburger B, but is also much larger and has more trimmings. By the intrinsic criteria, if you view both as being worth the price, then whatever your final choice, you at least get a good hamburger – you get your money's worth.

Dichotomies are the fundamental building blocks of everyday personal choices. Attached to virtually every nontrivial option are attributes that are desirable and attributes that are not desirable. To increase quality, one usually expects to pay more. To win a larger reward, one expects to take a greater risk. People are naturally wont to evaluate the upside versus the downside, the pros versus the cons, the pluses versus the minuses, the benefits versus the costs. One simply evaluates tradeoffs option by option – putting the gains and the losses on the balance to see which way it tips. The result of evaluating dichotomies in this way is that the benefits must be at least as great as the costs. In this sense, such evaluations provide a distinct notion of being good enough.

Definition 1.5
An option is **intrinsically rational** if the expected gains achieved by choosing it equal or exceed the expected losses, provided the gains and losses can be expressed in commensurable units.   □

Definition 1.6
An option is **intrinsically satisficing** if it is intrinsically rational.   □

By separating the positive (gain) and negative (loss) attributes of an option, I explicitly raise the issue of commensurability. It should be noted, however, that traditional utility theory also involves the issue of commensurability at least implicitly, since utility functions typically involve both benefits and costs, which are often summed or otherwise combined together to form a single utility function (for example, when forming a utility function for automobiles, positive attributes might be performance and reliability and negative attributes might be purchase and operating costs). Often such attributes can be expressed in, say, monetary units, but this is not always the case. Nevertheless, decision makers are usually able to formulate some rational notion of commensurability by appropriating or inventing a system of units. The issue was put succinctly by Hardin: "Comparing one good with another is, we usually say, impossible because goods are incommensurable. Incommensurables cannot be compared. Theoretically, this may be true; but in real life incommensurables *are* commensurable. Only a criterion of judgment and a system of weighing are needed" (Hardin, 1968, emphasis in original). Since my formulation of rationality requires explicit comparisons of attributes, the choice of units becomes a central issue and will be discussed in detail in subsequent chapters.

Intrinsic rationality is a weaker notion than substantive rationality, but it is more structured than procedural rationality. Whereas substantive rationality may be characterized as an attitude of "nothing but the best will do", and procedural rationality may be characterized as an attitude of "it has always worked before," intrinsic rationality may be characterized as an attitude of "getting what you pay for." Substantive rationality assures optimality but is rigid. Procedural rationality is efficient but amorphous. Intrinsic rationality is ameliorative and flexible. There can be only one substantively rational option (or an equivalence class of them) for a given optimality criterion, and there can be only one procedurally rational option for a given procedure,[8] but there can be several intrinsically rational options for a given satisficing criterion.

The quality of a substantively rational option will be superior to all alternatives, according to the criteria used to define it. The quality of a procedurally rational option may be difficult to assess, since no explicit criteria are required to define it. The quality of intrinsically rational options may be uneven, since options that provide little benefit but also little cost may be deemed satisficing. Thus, intrinsic satisficing can be quite different from satisficing *à la* Simon.

My justification for using the term "satisficing" is that it is consistent with the issue that motivated Simon's original usage of the term – to identify options that are good enough by directly comparing attributes of the options to a standard. This usage differs only in the standard used for comparison. Whereas Simon's standard is extrinsic (attributes are compared to an externally supplied aspiration level), my standard is intrinsic (the positive and negative attributes of each option are compared to each other). If minimum requirements are readily available, however, it is certainly possible to define satisficing in a way that conforms to Simon's original idea.

Definition 1.7
An option is **extrinsically satisficing** if it meets minimum standards that are already supplied.   □

Combining intrinsic and extrinsic satisficing is one way to remove some of the unevenness of intrinsic satisficing.

Definition 1.8
An option is **securely satisficing** if it is both intrinsically and extrinsically satisficing.
□

It will not be assumed that minimum standards can always be specified. But if they are, it will be assumed that they employ a rationale that is compatible with that used to define gains and losses. If minimum standards are not available, the decision maker must still attempt to evaluate the unevenness of intrinsically satisficing solutions.

---

[8] With heuristics such as satisficing *à la* Simon, however, there may be multiple options that satisfy an extrinsic satisficing criterion, and the agent need not terminate its search after finding only one of them.

This issue will be discussed in detail in Chapter 5. Throughout the remainder of this book, the term satisficing will refer solely to intrinsic satisficing unless stated otherwise. It will be assumed that gains and losses can be defined, and that these attributes can be expressed in units that permit comparisons.

1.3.2   Sociality

Competition, which is the instinct of selfishness, is another word for dissipation of energy, while combination is the secret of efficient production. (Edward Bellamy, *Looking Backward* (1888))

Self-interested human behavior is often considered to be an appropriate metaphor in the design of protocols for artificial decision-making systems. With such protocols, it is often taken for granted that each member of a community of decision makers will try

... to maximize its own good without concern for the global good. Such self-interest naturally prevails in negotiations among independent businesses or individuals... Therefore, the protocols must be designed using a *noncooperative, strategic* perspective: the main question is what social outcomes follow given a protocol which *guarantees that each agent's desired local strategy is best for that agent – and thus the agent will use it.* (Sandholm, 1999, pp. 201, 202; emphasis in original)

When artificial decision makers are designed to function in a non-adversarial environment, it is not obvious that it is either natural or necessary to restrict attention to noncooperative protocols. Decision makers who are exclusively focused on their own self-interest will be driven to compete with any other decision maker whose interests might possibly compromise their own. Certainly, conflict cannot be avoided in general, but conflict can just as easily lead to collaboration as to competition. Rather than head-to-head competition, Axelrod suggests that a superior approach is to look inward, rather than outward, and evaluate one's performance relative to one's own capabilities, rather than with respect to the performance of others.

Asking how well you are doing compared to how well the other player is doing is not a good standard unless your goal is to destroy the other player. In most situations, such a goal is impossible to achieve, or is likely to lead to such costly conflict as to be very dangerous to pursue. When you are not trying to destroy the other player, comparing your score with the other's score simply risks the development of self-destructive envy. A better standard of comparison is how well you are doing relative to how well someone else could be doing in your shoes. (Axelrod, 1984, p. 111)

This thesis is born out by the Axelrod Tournament (Axelrod, 1984), in which a number of game theorists were invited to participate in an iterated Prisoner's Dilemma[9]

---

[9] The Prisoner's Dilemma, to be discussed in detail in Section 8.1.3, involves two players who may either cooperate or defect. If one player cooperates and the other defects, the one who defects receives the best payoff while the one who cooperates receives the worst payoff. If both defect, they both receive the next-to-worst payoff, and if both cooperate, they both receive the next-to-best payoff (which is assumed to be better than the next-to-worst payoff).

tournament. The winning strategy was Rapoport's *tit-for-tat* rule: start by cooperating, then play what the other player played the previous round. What is interesting about this rule is that it always loses in head-to-head competition, yet wins the overall best average score in round-robin play. It succeeds by eliciting cooperation from the other players, rather than trying to defeat them.

Cooperation often involves *altruism*, or the notion that the benefit of others is one's ultimate goal. This notion is in contrast to *egoism*, which is the doctrine that the ultimate goal of every individual is to benefit only himself or herself. The issue of egoism versus altruism as an explanation for human behavior has captured the interest of many researchers (Sober and Wilson, 1998; Mansbridge, 1990a; Kohn, 1992). As expressed by Sober and Wilson:

Why does psychological egoism have such a grip on our self-conception? Does our everyday experience provide conclusive evidence that it is true? Has the science of psychology demonstrated that egoism is correct? Has Philosophy? All of these questions must be answered in the negative... The influence that psychological egoism exerts far outreaches the evidence that has been mustered on its behalf... Psychological egoism is hard to disprove, but it also is hard to prove. Even if a purely selfish explanation can be imagined for every act of helping, this doesn't mean that egoism is correct. After all, human behavior also is consistent with the contrary hypothesis – that some of our ultimate goals are altruistic. Psychologists have been working on this problem for decades and philosophers for centuries. The result, we believe, is an impasse – the problem of psychological egoism and altruism remains unsolved. (Sober and Wilson, 1998, pp. 2, 3)

Peirce, also, is skeptical of egoism as a viable explanation for human behavior:

Take, for example, the doctrine that man only acts selfishly – that is, from the consideration that acting in one way will afford him more pleasure than acting in another. This rests on no fact in the world, but it has had a wide acceptance as being the only reasonable theory. (Peirce, 1877)

It is not my intent to detail the arguments regarding egoism versus altruism as explanations for human behavior; such an endeavor is best left to psychologists and philosophers. But, if the issue is indeed an open question, then it would be prudent to refrain from relying exclusively on a rationality model based solely on self-interest when designing artificial entities that are to work harmoniously, and perhaps altruistically, with each other and with humans.

One of the possible justifications for adopting self-interest as a dominant paradigm for artificial decision-making systems is that it is a simple and convenient principle upon which to build a mathematically based theory. It allows the decision problem to be abstracted from its context and expressed in unambiguous mathematical language. With this language, utilities can be defined and calculus can be employed to facilitate the search for the optimal choice. The quintessential manifestation of this approach to decision making is von Neumann–Morgenstern game theory (von Neumann and Morgenstern, 1944). (See Appendix B for a brief summary of game theory basics.)

Under their view, game theory is built on one basic principle: individual self-interest – each player must maximize its own expected utility under the constraint that other players do likewise. For two-person zero-sum games (see Definition B.6 in Appendix B), individual self-interest is perhaps the only reasonable, non-vacuous principle – what one player wins, the other loses. Game theory insists, however, that this same principle applies to the general case. Thus, even in situations where there is the opportunity for group as well as individual interest, only individually rational actions are viable: if a joint (that is, for the group) solution is not individually rational for some decision maker, that self-interested decision maker would not be a party to such a joint action. This is a rigid stance for a decision maker to take, but game theory brooks no compromises that violate individual rationality.

Since many decision problems involve cooperative behavior, decision theorists are tempted to define notions of group preference as well as individual preference. The notion of group preference admits multiple interpretations. Shubik describes two, neither of which is entirely satisfactory to game theorists (in subsequent chapters I offer a third): "Group preferences may be regarded either as derived from individual preferences by some process of aggregation or as a direct attribute of the group itself" (Shubik, 1982, p. 109). Of course, not all group scenarios will admit a harmonious notion of group preference. It is hard to imagine a harmonious concept of group preference for zero-sum games, for example. But, when there are joint outcomes that are desirable for the group to obtain, the notion of group interest cannot be ignored.

One way to aggregate a group preference from individual preferences is to define a "social-welfare" function that provides a total ordering of the group's options. The fundamental issue is whether or not, given arbitrary preference orderings for each individual in a group, there always exists a way of combining these individual preference orderings to generate a consistent preference ordering for the group. In an landmark result, Arrow (1951) showed that no social-welfare function exists that satisfies a set of reasonable and desirable properties, each of which is consistent with the notion of self-interested rationality and the retention of individual autonomy (this theorem, known as Arrow's impossibility theorem, is discussed in more detail in Section 7.3).

The Pareto principle provides a concept of social welfare as a direct attribute of the group.

Definition 1.9

A joint (group) option is a **Pareto equilibrium** if no single decision maker, by changing its decision, can increase its level of satisfaction without lowering the satisfaction level of at least one other decision maker.   □

As Raiffa has noted, however, the Pareto equilibrium can be equivocal.

It seems reasonable, does it not, that the group *should* choose a Pareto-optimal act? Otherwise there would be alternative acts that at least some would prefer and no one would "disprefer". Not too long ago this principle seemed to me unassailable, the one solid cornerstone in an otherwise swampy area. I am not so sure now, and I find myself in that uncomfortable position in which the more I think the more confused I become.

One can argue that the group by its very existence should have a common bond of interest. If the members disagree on fundamentals (here, on probabilities and on utilities) they ought to thrash these out independently, arrive at a compromise probability distribution and a compromise utility function, and use these in the usual Bayesian manner. (Raiffa, 1968, p. 233, emphasis in original)

Adopting this latter view would require the group to behave as a *superplayer*, or, as Raiffa puts it, the "organization incarnate," who functions as a higher-level decision maker. Shubik refers to the practice of ascribing preferences to a group as a subtle "anthropomorphic trap" of making a shaky analogy between individual and group psychology. He argues that, "It may be meaningful . . . to say that a group 'chooses' or 'decides' something. It is rather less likely to be meaningful to say that the group 'wants' or 'prefers' something" (Shubik, 1982, p. 124). Shubik criticizes the view of the group as a superplayer capable of ascribing preferences according to some sort of group-level welfare function as being too narrow in scope to "contend with the pressures of individual and factional self-interest." Although Raiffa also rejects the notion of a superplayer, he still feels "a bit uncomfortable . . . somehow the group entity is more than the totality of its members" (Raiffa, 1968, p. 237).

Arrow expresses a similar discomfort: "All the writers from Bergson on agree on avoiding the notion of a social good not defined in terms of the values of individuals. But where Bergson seeks to locate social values in welfare judgments by individuals, I prefer to locate them in the actions taken by society through its rules for making social decisions" (Arrow, 1951, p. 106). Although Arrow does not tell us how such rules should be defined or, once defined, how they should be implemented, his statement nevertheless expresses the notion that societies may possess structure that is more complicated than can be expressed via individual values.

Perhaps the source of this discomfort is that, while individual rationality may be appropriate for environments of perfect competition, it loses much of its power in more general sociological settings. As Arrow noted, the use of the individual rationality paradigm is "ritualistic, not essential" (Arrow, 1986). What is essential, however, is that any useful model of society accommodate the various relationships that exist between the agents. But achieving this goal should not require artifices such as the aggregation of individual interests or the creation of a superplayer.[10] While such approaches may be recommended by some as ways to account for group interests, they may also manifest the limits of the substantive rationality paradigm.

Nevertheless, game theory, which relies exclusively upon self-interest, has been a great success story for economics and has served to validate the assumption of

---

[10] Margolis (1990) advocates a "dual-utilities" approach, comprising a social utility and a private utility, with the decision maker allocating resources to achieve a balance between the two utilities. Margolis' approach eschews the substantive rationality premise, and is very much in the same spirit as the approach I develop in subsequent chapters.

substantive rationality in many applications. This success, however, does not imply that self-interest is the only principle that will lead to credible models of economic behavior, it does not imply the impossibility of accommodating both group and individual interests in some meaningful way, and it does not imply that individual rationality is an appropriate principle upon which to base a theory of artificial decision-making entities.

Game theory provides a systematic way of analyzing behavior where the consequences of one player's actions depend on the actions taken by other players. Even single-agent decision problems can be viewed profitably as games against nature, for example. The most common solution concepts of game theory are dominance and Nash equilibria.

Definition 1.10

A joint option is a **dominant equilibrium** if each individual option is best for the corresponding player, no matter what options the other players choose.   □

Definition 1.11

A joint option is a **Nash equilibrium** if, were any single decision maker to change its decision, it would reduce its level of satisfaction.   □

A dominant equilibrium corresponds to the ideal situation of all players being able simultaneously to maximize their own satisfaction. This is a rare situation, even for games where coordination is possible. Nash equilibrium is a much more useful concept, but not all games possess pure (that is, non-random) Nash equilibria. Nash (1950) established, however, that if random play is permitted where each player makes decisions according to a probability rule (a mixed strategy), then at least one Nash equilibrium can be found for a finite-player, finite-action game.

In contrast to Pareto equilibria, Nash equilibria is a strictly selfish concept, hence is not amenable to cooperative play. But an individually rational player would have no incentive to agree to a Pareto equilibrium if that solution did not assure at least as much satisfaction as the player could be guaranteed of receiving were it to ignore completely the interests of the other players.

Definition 1.12

The minimum guaranteed benefit that a player can be assured of achieving is its **security level**.   □

Furthermore, a subgroup of players would have no incentive to agree to a joint solution unless the total benefit to the subgroup were at least as great as the minimum that could be guaranteed to the subgroup – its security level – if it acted as a unit (assuming transferable utilities which may be be reapportioned via side payments).

The **core** of an N-person game is the set of all solutions that are Pareto equilibria and at the same time provide each individual and each possible subgroup with at least their security levels (the concept of the core is discussed in more detail in Section 7.1). Unfortunately, the core is empty for many interesting and nontrivial games.

An empty core exposes the ultimate ramifications of a decision methodology based strictly on the maximization of individual expectations. There may be no way to meet all of the requirements that are imposed by strict adherence to the dictates of individual rationality. There are many ways to justify solutions that are not in the core, such as accounting for bargaining power based on what a decision maker calculates that it contributes to a coalition by joining it (e.g., the Shapley value), or by forming coalitions on the basis of no player having a justified objection against any other member of the coalition (e.g., the bargaining set).

I do not criticize the rationale for these refinements to the theory, nor do I criticize the various extra-game-theoretical considerations that may govern the formation of coalitions, such as friendship, habits, fairness, etc. I simply point out that to achieve a reasonable solution it may be necessary to go beyond the strict notion of maximizing individual expectations and employ ancillary assumptions that temper the attitudes and abilities of the decision makers. There are many such ingenious and insightful solution concepts but, as Shubik notes,

Each solution probes some particular aspect of rational individuals in mutual interaction. But all of them have had to make serious compromises. Inevitably, it seems, sharp predictions or prescriptions can only be had at the expense of severely specialized assumptions about the customs or institutions of the society being modeled. The many intuitively desirable properties that a solution ought to have, taken together, prove to be logically incompatible. (Shubik, 1982, p. 2)

This observation cuts to the heart of the situation: under von Neumann–Morgenstern game theory, any considerations of customs and peculiarities of the collective that are not explicitly modeled by the individual utility functions are extra-game-theoretic and must be accommodated by some sort of add-on logic. Much of the ingenuity and insight associated with game theory may lie in devising ways to force these considerations into the framework of individual rationality. While this practice may be appropriate for the *analysis* of human behavior, it is less appropriate for the *synthesis* of artificial decision-making entities, since any such idiosyncratic attributes must be an explicit part of the decision logic, not merely a *post factum* explanation for anomalous behavior. I suggest, however, that the problem is more fundamental than simply accounting for idiosyncrasies.

The critical issue, in my view, has to do with the structure of the utility functions. Before articulating this point, let me first briefly summarize utility theory as it is employed in mathematical games. Utility theory was developed as a mathematical way to encode individual preference orderings. It is built on a set of axioms that describe how a "rational man" would express his preference between two alternatives in a consistent

Table 1.1: Payoff array for a two-player game with two strategies each

	X_2	
X_1	s_{21}	s_{22}
s_{11}	$(\pi_1(s_{11}, s_{21}), \pi_2(s_{11}, s_{21}))$	$(\pi_1(s_{11}, s_{22}), \pi_2(s_{11}, s_{22}))$
s_{12}	$(\pi_1(s_{12}, s_{21}), \pi_2(s_{12}, s_{21}))$	$(\pi_1(s_{12}, s_{22}), \pi_2(s_{12}, s_{22}))$

way.[11] An expected utility function is a mathematical expression that is consistent with the preferences and conforms to the axioms. Since, in a game-theoretic context, an individual's preferences are generally dependent upon the payoffs (expected utilities) that obtain as a result of the individual's strategies and of the strategies available to others, an individual's expected utility function must be a function not only of the individual's own strategies, but of the strategies of all other individuals. For example, consider a game involving two players, denoted X_1 and X_2, such that each player has a strategy set consisting of two elements, that is, X_1's set of strategies is $S_1 = \{s_{11}, s_{12}\}$ and X_2's set of strategies is $S_2 = \{s_{21}, s_{22}\}$ (for this single-play game, strategies are synonymous with options). X_1's expected utility function would be a function $\pi_1(s_{1j}, s_{2k})$, $j, k = 1, 2$. Similarly, X_2's expected utility function is of the form $\pi_2(s_{1j}, s_{2k})$. Thus, each individual computes its expected utility as a function of both its own strategies and the strategies of the other players. These expected utilities may then be juxtaposed into a payoff array, and solution concepts may be devised to define equilibrium strategies, that is, strategies that are acceptable for all players. Table 1.1 illustrates the payoff array for a two-player game with two strategies each.

The important thing to note about this structure is that *it is not until the expected utilities are juxtaposed into an array so that the expected utility values for all players can be compared that the actual "game" aspects of the situation emerges.* It is the juxtaposition that reveals possibilities for conflict or coordination. These possibilities are not explicitly reflected in the individual expected utility functions by themselves. In other words, although the individual's expected utility is a function of other players' strategies, *it is not a function of other players' preferences.* This structure is completely consistent with exclusive self-interest, where all a player cares about is its personal benefit as a function of its own and other players' strategies, without any regard for the benefit to the others. Under this paradigm, the only way the preferences of others factor into an individual's decision-making deliberations is to constrain behavior to limit the amount of damage they can do to oneself. Pareto equilibria notwithstanding, a true notion of group rationality is not a logical consequence of individual rationality.

---

[11] This is not to say that the axioms cannot be generalized to deal with group preferences, but the theory has not been developed that way.

Table 1.2: Payoff matrix in ordinal form for the Battle of the Sexes game

	S	
H	D	B
D	(4, 3)	(2, 2)
B	(1, 1)	(3, 4)

Key: 4 = best; 3 = next best; 2 = next worst; 1 = worst

Luce and Raiffa summarize the situation succinctly:

> ...general game theory seems to be in part a sociological theory which does not include any sociological assumptions ... it may be too much to ask that any sociology be derived from the single assumption of individual rationality. (Luce and Raiffa, 1957, p. 196)

Often, the most articulate advocates of a theory are also its most insightful critics. Yet, such criticism is not often voiced, even by advocates of game theory as a model of human behavior. For example, consider the well-known Prisoner's Dilemma game (see Section 8.1.3). This game is of interest because possibilities for both cooperation and conflict are present, yet under the paradigm of individual rationality, only the joint conflict solution (the Nash equilibrium) is rational.

The Prisoner's Dilemma game may be an appropriate model of behavior when (a) the opportunity for exploitation exists, (b) cooperation, though possible, incurs great risk, and (c) defection, even though it offers diminished rewards, protects the participant from catastrophe. Many social situations, however, possess a strong cooperative flavor with very little incentive for exploitation. One prototypical game that captures this feature is the Battle of the Sexes game (Bacharach, 1976) to be discussed in detail in Section 8.1.2. This is a game involving a man and a woman who plan to meet in town for a social function. She (S) prefers to go to the ballet (B), while he (H) prefers the dog races (D). Each also prefers to be with the other, however, regardless of venue. The classical way to formulate this game is via a payoff matrix, as given in Table 1.2 in ordinal form, with the payoff pairs representing the benefits to H and S, respectively.

Rather than competing, these players wish to cooperate, but they must make their decisions without benefit of communication. Both players lose if they make different choices, but the choices are not all of equal value to the players. This game has two Nash equilibria, (D, D) and (B, B).

One of the perplexing aspects of this game is that it does not pay to be altruistic (deferring to the venue preferred by the other), since, if both participants did, they would each receive the worst outcome. Nor does it pay for both to be selfish (demanding the venue preferred by oneself) – that guarantees the next worst outcome for each player. The best and next-best outcomes obtain if one player is selfish and the other altruistic.

It seems that a way to account for the preferences of others when specifying one's own preferences would be helpful, but there is no obvious way to do this within the conventional structure.

Taylor (1987) addresses the issue of accounting for the interests of others by introducing a formal notion of altruism that involves transforming the game to a new game according to a utility array whose entries account for the payoffs to others as well as to oneself. Taylor suggests that the utility functions be expressed as a weighted average of the payoffs to oneself and to others. By adjusting the weights, a player is able to take into consideration the payoffs of others.

Taylor's form of altruism does not distinguish between the state of *actually relinquishing* one's own self-interest and the state of *being willing to relinquish* one's own self-interest under the appropriate circumstances. To relinquish unconditionally one's own self-interest is a condition of *categorical* altruism – a decision maker unconditionally modifies its preferences to accommodate the preferences of others. A purely altruistic player would completely replace its preferences with the preferences of others. A state of being willing to modify one's preferences to accommodate others if the need arises is a state of *situational* altruism. Here, a decision maker is willing to accommodate, at least to some degree, the preferences of others in lieu of its own preferences if doing so would actually benefit the other, but otherwise retains its own preferences intact and avoids needless sacrifice.

Categorical altruism may be too much to expect from a decision maker who has its own goals to pursue. However, the same decision maker may be willing to engage, at least to a limited degree, in a form of situational altruism. Whereas it is one thing for an individual to modify its behavior if it is sure that doing so will benefit another individual (situational), it is quite another thing for an individual to modify its behavior regardless of its effect on the other (categorical). In the Battle of the Sexes, If H knew that S had a very strong aversion to D (even though S would be willing to put up with those extremely unpleasant surroundings simply to be with H and thus receive her second-best payoff), H might then prefer B to D. But if S did not have a strong aversion to D then H would stick to his preference for D over B (in Section 8.1.2 I introduce situational altruism into this game).

This example seems to illustrate Arrow's claim that, when the assumption of perfect competition fails, "the very concept of [individual] rationality becomes threatened, because perceptions of others and, in particular, of their rationality become part of one's own rationality" (Arrow, 1986). Arrow has put his finger on a critical weakness of individual rationality: it does not provide a way to incorporate another's rationality into one's own rationality without seriously compromising one's own rationality. I do not assert that, under the theoretical framework of conventional game theory, it is impossible to formulate theoretical models of social behavior that go beyond individual interests and accommodate situationally altruistic tendencies while at the same time preserving individual preferences. However, the extant literature does not provide such a theory. I assert that it will be difficult to develop such a theory that remains compatible with the principle of individual rationality.

There are many ways to introduce categorical altruism into the design of artificial decision makers. One approach is to modify the decision maker's utility function to become a function of the group's payoff. In effect, the player is "brainwashed" into substituting group interests for its personal interests. Then, when acting according to its supposed self-interest, it is actually accommodating the group (Wolpert and Tumer, 2001). A somewhat similar, though less radical, approach is taken by Glass and Grosz (2000) and Cooper et al. (1996), who attempt to instill a social consciousness into agents, rewarding them for good social behavior by adjusting their utility functions with "brownie points" and "warm glow" utilities for doing the "right thing."

It is certainly possible for human altruists to interpret their sacrifice as, ultimately, a benefit to themselves for having made another's good their own (motivated, possibly, by such "pure" altruistic attributes as duty and love, or perhaps by "impure" altruistic attributes such as the sense of power that derives from having helped another (Mansbridge, 1990b)), but it seems less appropriate to ascribe such anthropomorphic interpretations (or motives) to artificial decision-making entities. While, granting that it is possible for a decision maker to suppress its own preferences in deference to others by redefining its own expected utility to be maximized, doing so is little more than a device for co-opting individual rationality into a form that can be interpreted as unselfish. Such a device only simulates attributes of cooperation, unselfishness, and altruism while maintaining a regime that is competitive, exploitive, and avaricious. Altruism, springing from whatever motive in man or machine, may often be accommodated in multi-agent relationships, but it does not follow that it can be accommodated within a regime that recognizes self-interest as the primary basis for rational decision making.

Social choice theory is another multi-agent formalism that has been widely studied. Like game theory, this theory has been developed largely on the foundation of individual rationality. For example, Harsanyi defines a social welfare function as a positive linear combination of individual utilities where each individual utility in this combination is a mapping of group options to individual utility. Each player then proceeds according to the substantively rational paradigm by maximizing its expected utility subject to any constraints that are relevant (Harsanyi, 1977).

The social welfare function modifies the decision maker's stance from a consideration of purely selfish preferences to a consideration of what are termed *moral* (or social) preferences, and gives weight to the interests of each participant. However, the sequence of mappings from group options to individual utilities and then from individual utilities to a group utility provides a very constrained linkage between one decision maker's preferences (for itself or for the group) and another decision maker's utilities and may not deal adequately with the rich diversity of interconnections that can exist in multi-agent groups. Furthermore, such mappings constitute unconditional (categorical) changes to the individual's utilities.

One of the characteristics of perhaps all societies, except for those that are either completely anarchic or completely dictatorial, is that group and individual preferences are woven together in a complex fabric that is virtually impossible to decompose into constituent pieces that function independently. Exclusive self-interest simply does not capture the richness and complexity of functional societies. On the other hand, to relinquish fundamental control over individual preferences and focus primarily on the preferences of the group as a whole may not be feasible, since individuals can be asked to make unreasonable sacrifices that place them in extremely disadvantageous situations. This suggests that functional societies must achieve some sort of equilibrium that is flexible enough to accommodate the preferences of both the individual and the group. Such an approach would be consistent with Levi's dictum that

> ... principles of coherent or consistent choice, belief, desire, etc. will have to be weak enough to accommodate a wide spectrum of potential changes in point of view. We may not be able to avoid some fixed principles, but they should be as weak as we can make them while still accommodating the demand for a systematic account. (Levi, 1997, p. 24)

Achieving, or at least approximating, equilibria involving both group and individual preferences is an essential condition for a system of autonomous artificial decision makers if they are to be representative of human groups. Obtaining such a state, however, requires a generalized notion of utility that seamlessly combines group and individual interests, even though it is individuals, and individuals only, who make the decisions. Such a utility theory must therefore be based on a notion of preference that allows group preferences to influence individual preferences and thereby to influence individual actions.

Accommodating group preferences must not leave the individual open to an unintentional or unacceptable degree of self-sacrifice. Thus, there must be a clear means of evaluation so that the individual can control the amount of compromise it is willing to consider. In other words, the individual must possess a means for self-control.

Heuristics offer no such capability. Under procedural rationality, once an individual adopts a rule that accommodates any form of compromise that exposes it to self-sacrifice, it becomes difficult to control the extent of its commitment without knowing beforehand the strategies of the other participants.

If one is willing to consider an option that is not strictly in its own best interest, one must be able to add some friction to the slippery slope of compromise. One way to do this is to adopt a satisficing stance, where satisficing is applied to the group as well as to the individual. Whereas optimization is strictly an individual concept, satisficing can be a social, as well as an individual, concept. For any group of decision makers, if the group and each of its members is willing to compromise sufficiently, there will exist a joint option that is good enough for the group as a whole and good enough for each member of the group according to their individual standards (this claim is made explicit in Section 7.2). This does not mean, of course, that the decision makers are obligated to accept this compromise option. It means only that it exists.

The remainder of this book explores the concept of intrinsic rationality, instantiated at both the individual and group levels, as a means of achieving an equilibrium of shared preferences and acceptable compromises. Intrinsic satisficing requires the specification of two general types of preferences – gains and losses. For a single-agent decision, it is conceptually straightforward to place each of the relevant attributes into one of these categories. When dealing with more than one decision maker, however, the interactions between them are not so readily categorized. Relationships are interconnected and conditional: one decision maker's gains and losses may affect other decision maker's gains and losses. Furthermore, the interconnections that exist between players must be at the level of preference interconnections, rather than action interconnections, as they are usually expressed in conventional game theory. The method of characterizing these preferences must be exhaustive, so that all possible relationships between decision makers can be represented, but at the same time it must be parsimonious, so that it is not more complex than it needs to be.

The central message of this book is that exclusive self-interest, coupled with strict optimality, is indeed an "excess of reasonableness." Self-interest is not the bedrock of rationality. Decision making, especially in group settings, can be ameliorated by relaxing the demands for optimization in its various forms (global maximization, constrained maximization, minimax, and even such "boundedly rational" approaches such as Simon's aspiration-level satisficing).

2 Locality

Order is not pressure which is imposed on society from without, but an equilibrium which is set up from within.

José Ortega y Gasset

Mirabeau: An Essay on the Nature of Statesmanship (Historical Conservation Society, Manila, 1975)

2.1 Localization concepts

Intrinsic rationality, as contrasted with substantive rationality, relies upon comparisons of attributes (gains versus losses) for each option rather then requiring the total ordering of preferences over all possible options. We may view intrinsic rationality as a *local information* concept, since only information pertaining to a particular option is involved in ordering the gain with respect to the loss. By contrast, we may view substantive rationality as a *global information* concept, since complete information regarding all options is required to form the rank ordering so that optimization can be performed.

One of the most successful concepts of science and engineering is the idea of localization. To localize a phenomenon is to delimit the extent of its influence. Some well-known examples of localization include: (i) *model localization*, such as lumped-parameter models that convert the partial differential equations of Maxwell's equations for modeling electromagnetic behavior into the ordinary differential equations of Kirchoff's laws; (ii) *spatial localization*, whereby a nonlinear dynamical system is constrained to operate near an equilibrium by confining inputs and initial conditions to be small enough to ensure that superposition approximately holds, thereby permitting the phenomenon to be characterized by a linear differential equation; (c) *temporal localization*, whereby a phenomenon is characterized over a small time interval, such as occurs with the design of a receding-horizon controller. These localizations are abstractions of global models that characterize behavior over the full extent of the problem. Their virtue is that they almost always require less information and computational capability than do their global counterparts and, consequently, can be implemented under conditions of limited information and computational resources. Their limitation is that they cannot be construed to be the best characterizations of the phenomena they purport to model and, hence, any notions of optimality based upon such models are difficult to justify.

Localization is also used in the assessment of interests. Substantive rationality is, fundamentally, itself a form of localization, since it concentrates on the concept of individual self-interest. We might call this *interest localization*. Just as the various concepts of localization in science and engineering have proven their practical indispensability, interest localization has become a dominant concept in models of multiple-agent decision making in economics, political science, and psychology. One of its virtues is its simplicity. It is the Occam's razor of interpersonal interaction, since it relies only upon the minimal assumption that an individual will put its own interests above everything and everyone else. It is understood, however, that this model is an abstraction of reality. Its value is that it provides insight into the workings of a complex society and can be used to explain past behavior or to predict future behavior.

Strict adherence to substantive rationality requires a total ordering of preferences and the capability to search this total ordering exhaustively if necessary (although hill-climbing techniques involving calculus and other efficient searching mechanisms may be employed). This results in an interesting paradox: strict compliance with interest localization (i.e., optimization) requires globalization of information (i.e., a total ordering). It also requires globalization of resources to insure a successful search.

The assumption that a decision maker possesses a total preference ordering that accounts for all possible combinations of choices for all agents under all conditions is a very strong assumption, particularly when the number of possible outcomes is large. In multi-agent decision scenarios, individuals may not be able to comprehend, or to even care about, a full understanding of their environment. They may be concerned mostly about issues that are closest to them, either temporally, spatially, or functionally. Thus, the preference orderings for an individual need not be of global extent, but may be restricted to proper subsets of the community or to proper subsets of conditions that may obtain. It may not be possible, and may not even be desirable, therefore, for each participant in a decision problem to possess a global (i.e., total) ordering that expresses its preferences for all possible combinations of choices for all agents in the community under all conditions. Substantive rationality, however, requires that these orderings be defined, and it is simply assumed that a rational decision maker either already possesses sufficient information about its environment to define a total ordering of all of its options or is able to learn these orderings as a result of experience.

One of the goals of this book is to demonstrate that, in terms of the scope of interest, intrinsic rationality includes the notion of group rationality and, in this sense, is actually more global than substantive rationality. It is, at the same time, more local in its ordering, since it requires intra-option comparisons (i.e., comparing different attributes of a given option) rather than inter-option comparisons (i.e., comparing attributes of one option to the same attributes of other options). Thus, the two forms of rationality offer a rather interesting parallel between localization and globalization of interest. Under the former, substantive rationality is the proper attitude and leads to the global (between option) orderings of the possibilities available to the decision maker. Under the latter, intrinsic rationality seems to be the appropriate attitude, and this notion of rationality leads to local (within option) orderings of the possibilities.

By giving up the insistence that individuals in a group must obtain their best outcomes, intrinsic rationality opens up a rational basis for the group as a whole, as well as for each individual, to achieve satisfactory results, provided that the participants, and perhaps the group as a whole as well, have flexible notions of what is satisfactory.

2.2   Group rationality

In Section 1.3.2 we described two interpretations of group preferences: (a) preferences that may be aggregated from individual preferences, such as social welfare functions, and (b) preferences that are direct attributes of the group itself, such as the preferences of a superplayer. Neither of these interpretations is completely acceptable to mainstream game theorists because neither is consistent with exclusive self-interest. They both violate the principle of substantive rationality.

As a first step in overcoming the seeming inconsistency between group and individual interests, let us replace substantive rationality with a less restrictive notion of rational behavior. Even after weakening the notion of rational behavior, however, it is not clear that either aggregating individual interests or creating a superplayer will permit reconciliation of group and individual interests. Thus, as a second step, let us explore new interpretations of group preferences.

Intrinsic rationality offers the possibility of a new notion of group preference, which I term *emergent group preferences*. To illustrate the type of decision problems that may be amenable to such a concept, consider the following example.

Example 2.1 The Pot-Luck Dinner Larry, Curly, and Moe are going to have a pot-luck dinner. Larry will bring either soup or salad, Curly will provide the main course, either beef, chicken, or pork, and Moe will furnish the dessert, either lemon custard pie or banana cream pie. The choices are to be made simultaneously and individually following a discussion of their preferences, which discussion yields the following results.

1. In terms of meal enjoyment, if Larry were to prefer soup, then Curly would prefer beef to chicken by a factor of two, and would also prefer chicken to pork by the same ratio. However, if Larry were to prefer salad, then Curly would be indifferent regarding the main course.
2. If Curly were to reject pork as being too expensive, then Moe would strongly prefer (in terms of meal enjoyment) lemon custard pie and Larry would be indifferent regarding soup or salad. If, however, Curly were to to reject beef as too expensive, then Larry would strongly prefer soup and Moe would be indifferent regarding dessert. Finally, if Curly were to reject chicken as too expensive, then both Larry and Moe would be indifferent with respect to their enjoyment preferences.

Larry, Curly, and Moe all wish to conserve cost but consider both cost and enjoyment to be equally important. Table 2.1 indicates the total cost (in stooge dollars) of each of the 12 possible meal combinations.

Table 2.1: The meal cost structure for the Pot-Luck Dinner

	Lemon custard pie	Banana cream pie
Beef (Soup/Salad)	23/25	27/29
Chicken (Soup/Salad)	22/24	26/28
Pork (Soup/Salad)	20/22	24/26

The decision problem facing the three participants is for each to decide independently what to bring to the meal. Obviously, each participant wants his own preferences honored, but no explicit notion of group preference is provided in the scenario. A distinctive feature of the preference specification for this example is that individual preferences are not even specified by the participants. Rather, the participants express their preferences as functions of other participants' preferences. Thus, they are not confining their interests solely to their own desires, but are taking into consideration the consequences that their possible options have on others. Such preferences are *conditional*. These interconnections between participants may imply some sort of group preference, but it is not clear what that might be. In fact, if the conditional and individual preferences are inconsistent, then there may be no harmonious group preference, and the group may be dysfunctional in the sense that meaningful cooperation is not possible. But if they are consistent, then some form of harmonious group preference may emerge from the conditional preferences (and any unconditional preferences, should they be provided). An important question is how we might elicit a group decision that accommodates an emergent group preference.

To formulate a von Neumann–Morgenstern game-theoretic solution to this decision problem, each participant must identify and quantify payoffs for every conceivable meal configuration that conform to their own preferences as well as give due deference to others. Notice that the unconditional preferences of each of the participants are not specified, nor are all of the conditional preferences specified. Unfortunately, substantive rationality makes it difficult to obviate such requirements. Thus, traditional game theory is not an appropriate solution methodology for this problem.

As an alternative we may consider a procedurally rational approach and formulate a heuristic rule for each participant. Since the decisions are to be made simultaneously, it is not possible for the participants to apply the conditional relationships, and there can be conflicting individual heuristics. Consider Larry. He has two possible rules from which to choose: soup or salad. If he were to assume that Curly would reject beef as too expensive (which would be a reasonable assumption, since beef is the most expensive of the three main course dishes), then choosing soup would be a legitimate heuristic. But, if Curly were to choose pork with Larry having preferred soup, Curly would then be stuck with his least favorite meal. He might therefore be unhappy, and a pie-throwing tantrum might ensue. Thus, Larry has a legitimate heuristic for choosing salad in deference to Curly's enjoyment.

The lack of a total ordering constraint in the problem statement presents serious problems to conventional game theory, since without this constraint it is impossible to impose standard solution concepts such as defining equilibria. The desire to apply traditional solution concepts such as game theory may motivate decision makers to manufacture orderings that that are not warranted. To solve this problem in a way that fully respects the problem statement, we need a solution concept that does not depend upon total orderings. It must, however, accommodate the fact that, even though agents may be primarily concerned with conditional local issues, these concerns can have wide-spread effects. In Chapter 6, a solution to the Pot-Luck Dinner problem is provided that is faithful to the problem statement and does not require the imposition of additional assumptions.

2.3 Conditioning

As we see with the Pot-Luck Dinner example, it is possible for one decision maker's preferences to be contingent upon the preferences of others. In fact, it is often far simpler to define payoffs in a context of the specific options available to others, rather than to attempt to define a global joint payoff.

To illustrate further, let us re-examine the Battle of the Sexes game, and suppose that, although H enjoys dog races, he is not a stereotypical machoistic male who has little consideration for the feelings of the opposite sex. Instead, let us cast him as a somewhat sensitive fellow who wants his friend to enjoy herself. He feels this way strongly enough to be willing to moderate his preference for the dog races if, but only if, S really hates that environment. He may express this feeling by defining two utility functions, one under the assumption that S detests the dog races, and the other under the assumption that she tolerates them. Such preferences are conditional for H, in that he does not commit to either preference independently of S's attitude. These utilities can be defined without H even knowing S's attitude about dog races. Notice, also, that it is possible for H to make these conditional evaluations without making direct reference to S's attitude about ballet.

Conditional preferences may be better understood by invoking a powerful analogy, namely, probability theory. Two of the most basic concepts of probability theory are the *law of total probability* and the *law of compound probability*. The law of total probability states that the probability of two disjoint events (events whose intersection is empty) is the sum of the individual probabilities, that is, if $A \cap B = \emptyset$,[1] then the probability of their union is the sum of the individual probabilities, that is, $P(A \cup B) = P(A) + P(B)$. The law of compound probability states that the probability of two non-disjoint events occurring simultaneously is the probability of one occurring, given that other does,

---

[1] See Appendix C for a definition of notation.

times the probability of the other one occurring, that is, $P(A \cap B) = P(A|B)P(B)$, where the expression $P(A|B)$ is the *conditional probability* of A given B. To illustrate these laws, let us look at a simple example. Let R be the event that it rains, let L be the event of light cloud cover, and let T be the event of thick cloud cover. Suppose that we learn from a reliable source (e.g., historical data) that the probabilities of rain, given the cloud cover conditions, are

$$P(R|L) = 0.1,$$
$$P(R|T) = 0.7.$$

These are conditional probabilities and indicate that there is a 10% chance of rain with light cover and a 70% chance of rain with thick cover. Conditional probabilities express local information in the sense that they characterize the chances of rain *given* specific weather conditions. These quantities can be defined regardless of the current weather conditions.

Now suppose you wish to plan an outdoor party next week. You examine the weather charts and determine that the probability of light cloud cover next week is 0.8 and that the probability of thick cloud cover next week is 0.2 (assuming that exactly one of these two conditions applies – this is an application of the law of total probability to the problem). According to the law of compound probability, the probability of simultaneous rain and light cloud cover is $P(RL) = P(R|L)P(L)$. The probability of simultaneous rain and thick cloud cover is $P(RT) = P(R|T)P(T)$. Finally, since the two weather conditions (RL) and (RT) are disjoint, we may apply the law of total probability to compute the probability of rain, namely,

$$P(R) = P(R|L)P(L) + P(R|T)P(T)$$
$$= 0.1 \times 0.8 + 0.7 \times 0.2$$
$$= 0.22,$$

that is, there is a 22% chance of rain. Thus, we see how to combine local, or conditional, information regarding the weather (that is, the probability of rain given specific cloud conditions) with the probability of the atmospheric conditions to obtain an assessment of the global, or unconditional, probability of rain.

Returning to the Battle of the Sexes game, let us use this same idea to evaluate H's preference for D. Let us suppose that H were to normalize his utility functions so that they have unit mass, that is, he has a unit of conditional utility to apportion among his options for each of S's possible states of mind. Suppose, given that S detests dogs, that H's conditional preference for D is 0.1, and, given that S tolerates dogs, his conditional preference for D is 0.7. He could then define two conditional utility functions as follows:

$$u_{H|S}(D|\neg D) = 0.1,$$
$$u_{H|S}(B|\neg D) = 0.9$$

(where $\neg$ is the negation symbol), to characterize his preferences given that S detests D, and

$$u_{H|S}(D|D) = 0.7,$$
$$u_{H|S}(B|D) = 0.3$$

to characterize his preferences given that S tolerates D.

Let us assume that by the time the moment of truth arrives, that is, the moment when H has to make a decision, he has somehow come into possession of S's attitude about D. He is now in a position to evaluate his preferences conditioned on that information. Suppose S's aversion to D is four times a strong as her tolerance for D. This attitude may be guaranteed by defining the normalized utility function u_S, which takes values $u_S(\neg D) = 0.8$ and $u_S(D) = 0.2$. Then H may compute his *conditioned* preference for D as

$$u_H(D) = u_{H|S}(D|\neg D)u_S(\neg D) + u_{H|S}(D|D)u_S(D)$$
$$= 0.1 \times 0.8 + 0.7 \times 0.2$$
$$= 0.22.$$

By a similar calculation, H may compute his conditioned preference for B as $u_H(B) = 0.78$. Thus, even though H strongly prefers D to B on his own, he lets his altruistic concerns dominate to the extent that, given S's attitude, he is willing to reverse his preferences.

Now let us consider the situation where S has a different attitude, such that her feelings about D are reversed, yielding $u_S(D) = 0.8$ and $u_S(\neg D) = 0.2$. Then we immediately obtain $u_H(D) = 0.58$, and H's altruistic tendencies are tempered by the fact that S would not be especially benefited by his sacrifice, and he does not reverse his preferences. It is important to appreciate the fact that this analysis may be done without needing to take into consideration S's attitude about B. In Section 8.1.2 we show in detail how this game may be formulated according to satisficing theory.

I have no proof that H's preferences should be assessed in the manner just cited. Indeed, it is possible to raise a number of objections to this way of evaluating preferences. One possible objection is the seemingly arbitrary normalization of preferences values. This objection may be superficially addressed by recalling that utility functions are, generally speaking, supposed to be invariant to origin and scale,[2] so normalizing them cannot change the preference orderings. There are circumstances where a deeper objection may be raised. Referring again to the Battle of the Sexes game, suppose H is willing to assign a value of 0.1, out of a possible unit of conditional utility, to the condition of attending D, given that S detests D, but is not willing to assign any positive conditional utility at all to the condition of attending B (that is, suppose H has an unmitigated aversion to B). Applying the law of total probability here would be

---

[2] In Chapter 5 we address in detail the invariance consequences of this normalization.

inappropriate, since the decision maker would not be willing to ascribe a high degree of desirability to the complement of D. The normalization structure requires, however, that he assign a value of 0.9 to this proposition, which implies that it is highly conditionally tolerable. In this situation, H's disposition seemingly permits him only to assign very low utilities to a proposition and its negation, but normalizing the utility functions does not permit him to withhold his commitment to apportion his entire unit of preferences among the options. This is exactly the problem raised by Shafer (1976) as a fundamental problem with the law of total probability – it does not permit agnosticism.[3] Likewise, normalizing utility functions does not permit abstention – the decision maker is obligated to apportion its entire unit of utility among the possibilities. A response to this possible criticism is that, while it is one thing to withhold belief, it is quite another to withhold action. If the purpose of making choices is ultimately to act, then the decision maker cannot simply do nothing (paralysis is itself an action with consequences), even if all alternatives are distasteful. Thus, if H really has a total aversion to B, he cannot entertain the idea of conditioning preference on its complement. He can ill afford to be altruistic with respect to attending a function to which he is completely averse. Instead, he must assign his entire unit of conditional utility to D, regardless of preferences held by S.

Another possible objection to the use of conditional preference relationships arises with the act of joining the product of preferences of one decision maker with preferences of another decision maker. So doing generates interpersonal utility dependencies that are typically frowned upon by conventional game theory. They are, however, allowed in social choice theory. In Chapter 5, I address the interpersonal comparisons issue in some depth, but for now let me plead guilty to making interpersonal comparisons while arguing that admitting them is an indispensable aspect of any decision theory that attempts to accommodate concerns that are wider than those of exclusive self-interest.

In defense of employing normalized conditional preference functions, let me offer the following observations. First, by normalizing the utilities (both conditional and unconditional) and combining them according to the law of compound probability, if we start with a unit of utility, we also end with a unit of utility, so utility combined this way exhibits conservation. Second, combining conditional utilities by means of weighted averages gives appropriate weight to all conditional possibilities, an intuitively pleasing interpretation. A third observation is that this structure represents a situational commitment, in that the one making the commitment is not required to follow through if the one to whom the commitment is made does not expect the commitment be honored. Finally, since this representation of preferences employs the mathematics

---

[3] The so-called Dempster–Shafer theory is an alternative to classical probability theory. It holds that probability theory cannot distinguish between uncertainty and ignorance, since it does not permit one to withhold belief from a proposition without ascribing the belief to its complement. Dempster–Shafer theory is an interesting alternative to probability theory, especially when subjective considerations dominate to the extent that the agent is simply unable to assign degrees of belief to all possibilities.

(but admittedly not the usual semantics) of probability theory, it offers a way to connect the interests of different agents in much the same way that probability theory extends from the univariate to the multivariate case. This feature will be a central issue in the formalization of satisficing games in Chapter 6 (also see Appendix D).

This last point may benefit from some elaboration. Multivariate probability theory is more than a simple extension of univariate probability theory. Except in cases of statistical independence, the joint distribution of two random variables cannot be obtained from the marginal distributions. Similarly, joint utilities between multiple decision makers cannot be obtained from the individual utilities unless the two decision makers have no consideration for each other. In other words, if the participants were all motivated by self-interest (every agent for itself), they would have no consideration for the interests of others and would not feel any obligation to them. Under such a regime, the only way a participant could accommodate the preferences of others would be to redefine its own unconditional preferences – it would have to "throw the game" via categorical altruism. We will examine the conditioning concept in much more detail in Chapter 6, where I present the idea of a satisficing game. For the present, however, the salient issue regarding conditioning is that it provides a mechanism for a decision maker to take into consideration the preferences of others when defining its own *conditioned* preferences *without categorically relinquishing its own original preferences*.

Before leaving the discussion of the structure of the utility functions, a brief comment regarding the issue of precision is in order. Suppose, with the Battle of the Sexes game, that H is not able to provide a single numerical value for S's attitude toward D, but is only able to specify an interval of values, (α_1, α_2) where $0 \leq \alpha_1 < \alpha_2 \leq 1$, such that $\alpha_1 \leq u_S(\neg D) \leq \alpha_2$. This lack of numerical precision (which also may be present with the conditional relationships) will create ambiguity in H, since his resulting conditioned preference will not be unique. This situation is very similar to the problem of dealing with imprecise probability measures. Fortunately, it is possible to extend the notions of precise probability measures to accommodate set-valued probabilistic relationships, and, by employing the mathematical structure of probability theory to characterize preferences, it is possible to construct conditioned set-valued utilities. I discuss this topic in more detail in Section 5.2.

2.4   Emergence

In addition to the temporal, or evolutionary, emergence that can occur with repeated play games such as iterated Prisoner's Dilemma, multi-agent systems may also exhibit a different emergence phenomenon, which we may call **spatial emergence**. Temporal emergence is an inter-game phenomenon that produces relationships between agents as time propagates, while spatial emergence is an intra-game phenomenon that produces relationships between agents as interests propagate through the group. A common example of spatial emergence occurs in both conventional game theory and social choice theory. This phenomenon corresponds to the *micro-to-macro*, or *bottom-up* view, with group preferences emerging as a consequence of individual interests. This approach often leads to the dilemmas and paradoxes that are so common in game theory. While such phenomena may serve as models for human behavior in certain circumstances, they may not lead to productive performance in the design of artificial autonomous systems. One way to circumvent such paradoxes is to adopt the *macro-to-micro*, or *top-down* approach to spatial emergence, where the interests of the group as a whole are paramount, and individual preferences are imposed as consequences of group interest as specified by a superplayer. This latter approach, however, requires very restrictive assumptions about the group and may not be appropriate if decision making is distributed among a number of autonomous agents.

Neither the top-down nor bottom-up notion of spatial emergence is a natural fit to the Pot-Luck Dinner example. A bottom-up approach to this problem would to require that each participant specify his own preferences for each possible meal and then work from there to a social choice that would be acceptable to the entire group. But the sufficient information by which to do this is not contained in the problem statement. On the other hand, a top-down approach would require the specification of some group good, such as avoiding conflict, and each participant would have to make a choice that would be consistent with that good. Such a good, however, is likewise not provided by the problem statement. Thus, both top-down and bottom-up emergence require the provision of additional information or conditions.

The Pot-Luck Dinner example, however, is characterized by local conditional interdependencies; neither individual nor group orderings of preferences are exhaustively provided by the problem statement. The example seems to call for a new viewpoint, which I characterize as an *inside-out*, or *meso-to-micro/macro*, view, where intermediate-level conditional preferences propagate up to the group level and down to the individual level. With this model, both individual and group behavior may emerge as consequences of local conditional interests that propagate throughout the group. (In Section 6.4 I show how this emergence may occur.)

If antagonism exists between members of the group, a harmonious group preference may not emerge. However, if the members of the group are willing to compromise, it may still be possible to define a group preference which, although it does not represent ideal cooperation, does at least offer a weaker notion of acceptability as an alternative to the total failure of the group to function. Thus, it is important for the individual decision makers to possess an ability to compromise in the interest of both their own welfare and the welfare of others.

This book is largely an outgrowth of attempts to understand how one might design an artificial social system that accommodates bottom-up, top-down, and inside-out views. As technology continues to advance, the demand for the creation of autonomous decision-making systems that function in communal environments is certain to increase.

One of the questions that must be addressed deals with the "attitude" that will be built into these systems. If they are designed from the perspective of interest localization, they may be naturally competitive and approach social interaction with an aggressive posture. On the other hand, if they are designed from a broader perspective, they may be more amenable to coordinative and cooperative behavior. They may be willing to compromise, even to sacrifice some self-interest for the good of others.

2.5   Less is more

Substantive rationality has unquestionably been the dominant paradigm for decision making for more than six decades. It provides unparalleled respectability, confidence, and security. Substantive rationality is appropriate when well-defined mathematical models are available to characterize performance, cost, and uncertainty. It is a *superlative* concept – it yields the best, but only the best, solution.

Procedural rationality is a viable alternative to substantive rationality when mathematical models are either not available or are only very approximate. In such circumstances, the procedural claim to rationality is to follow a set of rules that have been formulated by an authority. Procedurally rational decisions differ fundamentally from substantively rational solutions in that they are obtained without relying on comparisons to the quality of other possible solutions. Procedural rationality is a *positive* concept – it yields solutions that have worked in the past under similar circumstances.

Between the superlative grammatical degree of being best and the positive grammatical degree of being good, there is the *comparative* grammatical degree of being better. The adjective *better* requires binary comparisons between valuations. Such comparisons are local, in that they involve only the relationship of one valuation to another, in contrast to the notion of bestness, which requires a global comparison of one valuation with all others. From this point of view, we may classify an option as being "better" if, upon forming the binary comparison of gains versus losses, the gains exceed the losses. Remember that this is an intra-option comparison, rather than an inter-option comparison and, in this sense, we may claim that intrinsic rationality and satisficing achieve better, that is, ameliorative, results.

The three degrees of comparison provide significantly different ways to evaluate choices. The positive degree does not permit comparisons with anything.[4] Under this paradigm, there is no way to determine a degree, or grade, of goodness. An option either is, or is not, good. The superlative degree requires comparisons of each option with every other option. Under this paradigm, there is no way to determine a degree, or grade, of bestness. An option either is, or is not, the best. The superlative degree not only invites, but indeed demands, head-to-head competition. The comparative degree, as I

---

[4] *Post factum* comparisons may be used to adjust future behavior, but have no effect on the current decision.

develop the notion, requires comparisons of attributes associated with a given option but does not employ comparisons with other options. Under this paradigm, it is possible to determine degrees, or grades, of betterness – not with respect to other options, but with respect to the intrinsic qualities of the specific option.

The degree of betterness can be defined in terms of the size of the gains over the losses. This notion of comparison does not involve competition between options. It is not exclusive, since many options may qualify as being ameliorative.

Given an arbitrary decision problem, there is always a strong desire to adapt it, if possible, to the formalisms of optimization. The main argument for doing so is that otherwise the decision maker would needlessly sacrifice either cost or performance. It is assumed from this point of view that only individually rational decision mechanisms are viable. Equilibrium, by this model, is a condition whereby a decision maker is assured of maximizing the minimum benefit to itself, even if this means minimizing the maximum benefit to others. An intransigent self-interested maximizer lives a pessimistic existence in a world where all others are viewed, ultimately, as adversaries.

Such an environment is not necessarily natural for artificial systems that are required to function coordinatively. For such systems, optimality, though desirable, is subordinate to functionality. Individual interests must be accommodated, but so must group interests. Consider the following scenarios:

Example 2.2 Factory scheduling. A factory consists of a number of autonomous work sectors, each of which must produce some item. Each position consumes varying quantities of resources (energy, materials, labor, etc.), which are all drawn from a common and limited store. Each sector's goal is to maintain its profitability by producing as much as it can. The corporate goal is to maintain overall profitability, but there is no central control – each sector must make its own decision.

Example 2.3 Transporting a bulky object. A group of movers (which may be mobile robots) must transport a bulky object from point A to point B without dropping it. Each member of this group wishes to avoid hazards (e.g, clutter, obstacles, etc.) as much as possible. There is no leader–follower hierarchy. Each individual must make its own decision.

In the factory scheduling problem viewed from the point of view of substantive rationality, each sector is dominated by self-interest – producing as much of its product as possible – and views the other stations as competitors. Any cooperation between the sectors is purely incidental. The self-interested optimal solution would be a Nash equilibrium point such that, if any sector changed its demand for resources, it would either reduce its output or violate the resource constraints. Unfortunately, there may be many such equilibria, and there is no way to choose definitively between them without imposing some notion in addition to strict self-interest. The social welfare solution would be to choose an equilibrium that maximizes the weighted sum of the payoffs to all individual utilities. This practice is arbitrary in that there is no guarantee that a set of options that maximize corporate (social) welfare will satisfy the individual decision makers. For example, corporate benefit may be maximized by focusing all resources on one highly profitable sector, but such a stance is not likely to be an acceptable joint decision among a larger collection of self-interested utility maximizers.

The object-transporting scenario is a situation where the decision makers clearly must subordinate individual welfare to group welfare. That is, except for the desire to avoid hazards, each individual's self-interest is served by the group's interest. One way to view this situation under substantive rationality is as a constrained maximization problem, where the participants strive to maximize their individual utilities subject to the constraint that the goal is achieved. Here again, however, individual self-interest necessitates that, ultimately, each participant must view the others as a competitor for resources (i.e., avoiding hazards), and an optimal solution will be one such that, if any participant were to deviate from it, it would either increase its exposure to hazard or violate the group success constraint.

Self-interest is a very simple concept. Also, as these examples suggest, it is a very limiting concept, since it justifies ignoring the preferences of others when ordering one's own preferences. The advantage of invoking exclusive self-interest to define the solution is that doing so may drastically reduce the complexity of a model of the group of interest. The price for doing so is the risk of compromising group interests when individual preferences dominate or of distorting the real preferences of the individuals when group interests dominate. The root of the problem, in both of these cases, is the lack of a way to account for both group and individual interests in a seamless, consistent way.

Strict self-interest can create scarcity. Scarcity can come about in two ways, as illustrated in Example 2.5. Since all sectors must draw resources from a common source, there simply may not be enough to go around, and all of the participants may consequently suffer. This is the "tragedy of the commons" described by Hardin (1968). This kind of scarcity is unavoidable and emerges from the structure of the problem. A second form of scarcity is also manifest in this problem. Suppose there is a considerable amount of resource, but one sector has a superior ability to accumulate it and is able to obtain as much as it wants. Under substantive rationality, it will do so – it has no choice – thereby creating a condition of scarcity for the less capable sectors. Perhaps such pronounced iniquities could be prevented by imposing regulations, but regulations do not explicitly account for group interests; they merely limit individual interests. Regulations are arbitrary and can lead to the inefficient distribution of resources.

It is an objective fact that the goal of trying to be better than someone else is different from the goal of trying to do a job well.[5] When a person devotes energy to beating another person, he or she must expend energy to thwart the other – energy that could be used for other, more constructive purposes. The same is true with artificial decision makers. Returning to the factory example, it is obvious that, were a powerful decision

---

[5] Recall the bumper sticker slogan: Whoever has the most toys when he dies, wins.

Table 2.2: Frameworks for decision making

Rationality concept	Decision paradigm	Solution concept	Knowledge requirements
Substantive (extrinsic) rationality	Superlative degree (optimal)	Maximal expectations	Global mathematical models
Intrinsic rationality	Comparative degree (dichotomous)	Acceptable tradeoffs	Local mathematical models
Procedural rationality	Positive degree (heuristic)	Authoritative procedures	Local behavioral rules

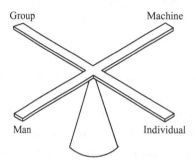

Group   Machine

Man   Individual

Figure 2.1: Achieving sociological and ecological balance.

maker to accommodate the desires of others or of the group in general, it would feel some pressure to police itself and limit its consumption for the benefit of others, even though this would mean achieving less than what is possible.

In light of the obvious wisdom of sharing, it does not seem appropriate to insist on a paradigm of exclusive self-interest when designing coodinative artificial decision-making systems. Even if the systems are designed by different parties and respond to their different interests, they may still accommodate each other.

In suggesting a relaxation of substantive rationality, I do not advocate abandonment of utility theory or the adoption of purely heuristic approaches. Rather, I advocate the consideration of a third alternative for rationality, one that fills the lacuna between the substantive and the positive notions. Table 2.2 illustrates this approach to decision-making and compares its features with those of substantive rationality and procedural rationality.

A **social contract** is, essentially, the set of all coordinating regularities by which a society operates, *as if* in consequence of an agreement, due to a general preference to conform to that regularity. No member of the society is bound to such a contract, but members of the society generally sustain it because it is in their interest to do so (Lewis, 1969). When designing artificial decision makers that must function in a group to achieve cooperative behavior, the designer would be wise to conform the design to a social contract that not only tolerates, but promotes, cooperation. The social contract implicit with von Neumann–Morgenstern game theory is that each individual will do the best it can for itself, regardless of the consequences to others, under the assumption that others will do likewise. Such a contract may be too narrow in scope to accommodate a society of artificial decision makers that must cooperate. The notion of satisficing as developed in this book, however, may provide the basis for the design of artificial societies that are able to achieve a sociological balance between individual and group interests.

Furthermore, in complex societies involving both humans and machines, it is important that these two components function in harmony. There is a growing body of sociological theory that human behavior is much too complex to be explained completely by exclusive self-interest (Sober and Wilson, 1998; Mansbridge, 1990a). To the extent that human societies conform to the satisficing concept, the design of artificial decision-making entities under the satisficing paradigm will also be compatible with human behavior. This suggests that the effective design of artificial societies that are to cooperate among themselves and with humans must be both sociologically balanced between group and individual interests and ecologically balanced between the artificial society and human society. Figure 2.1 illustrates this two-way balancing concept.

3 Praxeology

A mathematical formalism may be operated in ever new, uncovenanted ways, and force on our hesitant minds the expression of a novel conception.

Michael Polanyi

Personal Knowledge (University of Chicago Press, 1962)

The basic principle of decision making based on substantive rationality is very simple: one seeks to maximize expected utility. This principle has led to a body of mathematics that accommodates ways to rank-order expectations and to search or to solve for the option (or options) that meet the optimality criteria. The major mathematical components of this approach are utility theory, probability theory, and calculus.

The basic principle of decision making based on intrinsic rationality is also very simple: one seeks acceptable tradeoffs. To be useful, this principle must be supported by a body of mathematics that accommodates ways to formulate tradeoffs and to identify the options that meet the satisficing criteria. This chapter introduces such a mathematical structure. It also is composed of utility theory, probability theory, and calculus, but with some important differences in the structure and the application of these components (Stirling and Morrell, 1991; Stirling, 1993; Stirling, 1994; Stirling et al., 1996a; Stirling et al., 1996c; Stirling et al., 1996b; Goodrich et al., 1999; Stirling and Goodrich, 1999a; Goodrich et al., 2000).

3.1 Dichotomies

In terms of grammar, there is a logical gap between the superlative paradigm of being best and the positive paradigm of being merely good. This gap is filled by the *ameliorative* paradigm of being better. Rank-ordering is an extrinsic exercise involving inter-option evaluations; that is, comparing a given attribute of one option to the same attribute of another option or to a fixed standard. But this is not the only way to evaluate options. We may also make intra-option evaluations by forming dichotomies involving different attributes of a given option.

The approach I advocate is very simple and precedes formal theories regarding its use: a common way people evaluate personal and business options is to compare potential gains with potential losses. Benjamin Franklin did it this way:

> [M]y way is, to divide half a Sheet of Paper by a Line into two Columns, writing over the one Pro, and over the other Con. Then during three or four Days Consideration I put down under the different Heads short Hints of the different Motives that at different Times occur to me for or against the Measure. When I have thus got them all together in one View, I endeavor to estimate their respective Weights; and where I find two, one on each side, that seem equal, I strike them both out: If I find a Reason pro equal to some two Reasons con, I strike out the three. If I judge some two Reasons con equal to some three Reasons pro, I strike out the five; and thus proceeding I find at length where the Ballance lies; and if after a Day or two of farther Consideration nothing new that is of Importance occurs on either side, I come to a Determination accordingly. And tho' the Weight of Reasons cannot be taken with the Precision of Algebraic Quantities, yet when each is thus considered separately and comparatively, and the whole lies before me, I think I can judge better, and am less likely to make a rash Step; and in fact I have found great Advantage from this kind of Equation, in what may be called Moral or Prudential Algebra. (Franklin, 1987, p. 878)

As a formalized means of decision making, dichotomous evaluations appear in at least two very different contexts: economics and epistemology. The former is practical and concrete, the latter is theoretical and abstract.

3.1.1 Cost–benefit analysis

Economists implemented the formal practice of cost–benefit analysis in the 1930s to evaluate the wisdom of implementing flood control policies. The usual economics procedure is to express all costs and benefits in monetary units and to endorse an option if "the benefits to whomsoever they accrue are in excess of the estimated costs" (Pearce, 1983). An objection to cost–benefit analysis is that Pareto equilibria cannot be guaranteed, since the procedure may commend an option by which the welfare of some participants is improved, even though that of others might be reduced. This criticism is mitigated by the argument that, since the utility (money) is transferable, those who are made better off can compensate the others by means of side payments after the option is implemented and still remain better off (Kaldor, 1939; Hicks, 1939). By so doing, however, it is possible that the distribution of income may change in a way that reverses the original decision to undertake the project, and the same cost–benefit analysis may encourage the undoing of the project it had previously endorsed (Scitovsky, 1941).

Consider the following two-agent decision problem. X_1 and X_2 consider making a transaction with the following cost–benefit structure (in dollars):

	Cost	Benefit
X_1	8	5
X_2	−10	5

Notice that the cost for X_2 is negative, that is, X_2 is actually paid a bonus for simply making the transaction, independent of the results that accrue as a result, whereas X_1 must pay a fee to make the transaction. Since the sum of the benefits is greater than the sum of the costs, this transaction is acceptable under the simple rules of cost–benefit analysis. Unfortunately, this result as it stands is not a Pareto equilibrium, since X_1 is made worse off as a result. Now let us suppose that X_2 makes a side payment of $9 to X_1, after which the net gain to each participant is $6, and all is apparently well. But let's pursue the analysis one step further and suppose that now, after this redistribution of wealth, the two decision makers consider reversing their transaction by paying the same costs but negating the benefits. The actual out-of-pocket costs, however, are adjusted by the net increases they received with the first transaction, resulting in a revised cost–benefit structure of the form

	Cost	Benefit
X_1	2	−5
X_2	−16	−5

Since the revised costs (−$14) are less than the benefits (−$10), it is apparent that reversing the original transaction is actually acceptable under cost–benefit analysis. This example admittedly is a bit contrived, since it depends on the usage of negative costs, but this is not prohibited under the simple rules of cost–benefit analysis. Such a result obtains because the interests of both entities are aggregated into a single monolithic interest by comparing the total costs with the total benefits.

Despite its flaws, cost–benefit analysis, founded as it is on a very simple principle, has proven to be a useful way to reduce a complex problem to a simpler, more manageable one. One of its chief virtues is its fundamental simplicity. Since it accounts primarily for group interests, however, traditional cost–benefit analysis is not powerful enough to produce decisions that are rational from both group and individual perspectives.

3.1.2   Epistemic utility theory

Epistemic logic deals with the classification of propositions on the basis of knowledge and belief regarding their content. A major school of epistemology defines knowledge as true justified belief and takes as its major challenge the activity of justification. The goal of this line of thinking is ultimately to converge on the truth and nothing but the truth. As James observed, however, there is more than one way to approach the quest for knowledge:

> There are two ways of looking at our duty in the matter of opinion – ways entirely different, and yet ways about whose difference the theory of knowledge seems hitherto to have shown very little concern. We must know the truth, and we must avoid error – these are our first and great commandments as would-be knowers; but they are not two ways of stating an identical commandment, they are two separable laws . . .

> Believe truth! Shun error! – these, we see, are two materially different laws; and by choosing between them we may end by coloring differently our whole intellectual life. We may regard the chase for truth as paramount, and the avoidance of error as secondary; or we may, on the other hand, treat the avoidance of error as more imperative, and let truth take its chance. (James, 1956, pp. 16–17)

The view that the chase for truth is paramount is consistent with substantive rationality: demand the best (in terms of maximizing expectations) and nothing less than the best solution. To fashion a view consistent with the notion that the avoidance of error is imperative, however, we require a system of decision making that is fundamentally different from the standpoint of substantive rationality. This latter view actually requires something stronger – it requires a modification to the concept of knowledge. In the opening lines of his book, *The Enterprise of Knowledge*, Levi offers the following insight:

> Knowledge is widely taken to be a matter of pedigree. To qualify as knowledge, beliefs must be both true and justified. Sometimes justification is alleged to require tracing of the biological, psychological, or social causes of belief to legitimating sources. Another view denies that causal antecedents are crucial. Beliefs become knowledge only if they can be derived from impeccable first premises according to equally noble first principles. But whether pedigree is traced to origins or fundamental reasons, centuries of criticism suggest that our beliefs are born on the wrong side of the blanket. There are no immaculate preconceptions.
>
> When all origins are dark, preoccupation with pedigree is self-defeating. We ought to look forward rather than backward and avoid fixations on origins.
>
> Epistemologists should heed similar advice. Whatever its origins, human knowledge is subject to change. In scientific inquiry, men seek to change it for the better. Epistemologists ought to care for the improvement of knowledge rather than its pedigree. (Levi, 1980, p. 1)

Levi's challenge is a natural sequel to and is compatible with James' assertion that focusing on the avoidance of error is fundamentally different from focusing on seeking the truth. Whereas the conventional view is to seek truth at all costs, Levi is content with the stance of amelioration – change for the better. By relaxing the exclusive demand for truth, he thereby opens the way for the avoidance of error to receive increased consideration. He proposes a theory of knowledge dynamics, called epistemic utility theory, to pursue this goal. The fundamental basis of this theory involves the evaluation of carefully constructed dichotomies.

3.2   Abduction

One who would follow James' suggestion that we treat "the avoidance of error as more imperative" would assume a cautious stance and would be reluctant to incur significant risk of making a mistake. An argument can be advanced that error-avoidance is nothing more than cautious truth-seeking. But if James is right in his assertion that truth-seeking and error-avoidance "are two separable laws," then each "law" should admit its own distinct philosophy of decision making. Let us consider the effect of such a change of emphasis.

Minimizing the probability of error it not equivalent to avoiding error. Indeed, if an expressed aim of our inquiry is to avoid error, we may comply with this aim simply by refusing to make any choice at all. Though this course of action is faithful to the injunction, it leads to a vacuous decision. Evidently, the decision maker must be willing to incur some risk of error if a meaningful decision is to be made. But if any risk of error is to be incurred, it would seem that one could hardly improve on the concept of minimizing the probability of error.

There is, however, another notion of error-avoidance that is distinct from this standard approach. Suppose some measure of economic, political, moral, cognitive, aesthetic, or personal value could be ascribed to a proposition if it were adopted, independently of its truth or error. If a proposition were to supply valuable information[1] of these types, the decision maker should be more willing to risk error to adopt it than if it supplied information of only marginal value. Such considerations, though having no direct bearing on truth, represent an indispensable component of the decision maker's goals and should not be ignored when evaluating propositions. As Levi puts it about truth-seekers, "The choice of a conclusion to a given question on given evidence ought ideally to satisfy two desiderata: the answer chosen ought to be true, and the answer ought to supply information of the sort demanded by the question." (Levi, 1984, p. 52). If the decision maker stands to risk error as a result of making a choice, the choice must be informative as well as plausible. Popper expressed a similar concern: "Yet we must also stress that *truth is not the only aim of science.* We want more than truth: what we look for is *interesting truth . . .* " (Popper, 1963, p. 229, emphasis in original).

An error-avoider thus imposes an additional restraint on decision making that a naive truth-seeker does not require; namely, an aversion to committing to propositions that do not supply information, even if they are true. As a result, an error-avoider pursues a goal that is distinct from the goal pursued by a truth-seeker. Whereas a truth-seeker concentrates on searching for the truth with no independent conditions regarding its informational value,[2] an error-avoider concentrates on avoiding propositions that are either likely to be in error or uninformative or both. Given whatever evidence is available, it may be possible for an error-avoider to eliminate some propositions, leaving only those which, on the basis of the available knowledge, may be deemed to be reasonable as measured by their absence of error and their information value. As additional knowledge becomes available, more and more propositions may be eliminated. In the limit, as all relevant knowledge is obtained, the ideal error-avoider should eliminate all

---

[1] Information, as used in this context, does not correspond to either Shannon information (a concept defined in terms of entropy) or Fisher information (a concept defined in terms of statistical variance). Rather, it is a consideration of the importance, potential usefulness, or significance of a proposition or set of propositions.

[2] Informational value, as used in this context, is distinct from the notion of "value of information" of conventional decision theory (Raiffa, 1968), which deals with the change in expected utility that obtains if uncertainty is reduced or eliminated from a decision problem.

but the truth. If this situation were to obtain, then truth-seeking would be the limiting case of error-avoidance.

Informational value and truth value are the essential components of the dichotomy that Levi is constructing. To make this dichotomy explicit, we must formulate precise mathematical expressions for these concepts. The following example motivates our discussion.

Example 3.1 Theories of motion. Suppose we are to choose between rival theories to explain a given phenomenon, such as the observed fact that heavy things, such as stones, fall when released, and light things, such as smoke, rise when released.

Aristotle proposed the theory that heavy objects fall because the earth is their natural place, and light objects rise because the sky is their natural place. According to Aristotle, heavy objects possess a quality called "gravity," which cause them to descend, and light objects possess a property called "levity," which cause them to rise. In his view, these natural motions require no force. "Violent motions," under this view, are characterized by the fact that force is required to effect them, such as with a stone, which naturally would descend unless force were applied to lift it.

Newton also proposed a theory of motion. His basic premise is that the natural motion of an object is uniform motion in a straight line in a global, inertial, Euclidean reference frame. His description was much more sophisticated than Aristotle's: the law of universal gravitation. "Violent motion," for Newton, is acceleration, which requires force to change the motion from natural, uniform straight-line motion to some other type of motion.

Einstein proposed a third theory: the natural motion of an object is a straight line relative to a *local* inertial reference frame in curved spacetime. The action-at-a-distance concept that is central to the law of universal gravitation is replaced with the concept that an object simply follows its own world line.

Let us first consider the theories described in the above example in terms of truth support without regard for information content. They differ substantially regarding the descriptive question: "How do things move?" A powerful way to characterize truth support for propositions such as these is through probability theory. Through experimentation or personal convictions, a decision maker may arrive at a probability or family of probabilities that describe the level of belief associated with these three propositions. Whether or not these probabilities are numerically precise is not important here. What is important is that there be some means of assessing the degree of belief about the truth of these theories.

This probabilistic model of characterizing truth support is widely accepted, although various alternatives exist, such as Dempster–Shafer theory, possibility theory, and fuzzy set theory. In this book I do not wish to focus on the appropriateness or relative merits of these various methods of expressing belief. I adopt probability theory as the model for truth support, but recognize that the resulting decision theory might also be developed according to any of the alternative models for representing truth support.

It also becomes instructive to evaluate these three theories of motion strictly in terms of information. Such a consideration is termed an **abductive** inference (Levi, 1980). The process of abduction, or appealing to the best explanation, is the process of evaluating propositions in terms of how well they meet the demand for information. One makes an abductive inference when one evaluates a potential answer to a question in terms of how valuable the answer is if we set aside all considerations of its truth. For example, issues such as simplicity, testability, explanatory power, and predictive power could properly be considered by a scientist in making abductive judgments about a theory. Abduction guides the formation of propositions for serious consideration. Whitehead recognized the special significance of the abductive step:

> It is more important that a proposition be interesting than that it be true. This statement is almost a tautology. For the energy of operation of a proposition in an occasion of experience is its interest, and is its importance. But of course a true proposition is more apt to be interesting than a false one. (Whitehead, 1937, Part 4, Chapter 16)

Among the consequences of adopting one of the rival theories of motion is that the decision maker must appeal to the theory to explain and predict observed and future behavior. In terms of abductive considerations, therefore, a modern person will likely glean very little information from Aristotle's description; it simply is not very useful. It cannot be used, say, to predict how long it will take for a dropped stone to hit the ground. Both Newton's and Einstein's theories, however, are able to make such predictions – they are thus more interesting. Also, Newton's action-at-a-distance model presents problems that Einstein's model addresses. Thus, when considering these theories in light of information only, a scientist could form an abductive ranking of them in terms of this consideration, independently of truth. However, an abductive "conclusion" is not a decision leading to action. Rather, it is simply a determination to retain a proposition for further evaluation, based on our interest in the proposition, rather than on its truth. Consider the following problem as a further example of informational value:

Example 3.2 Melba receives an encoded message from Milo, but there is ambiguity in the decoding, so that Melba knows only that one of three messages was sent:
M-1: Would you like a piece of cheese?
M-2: Would you like to go to a ball game?
M-3: Will you marry me?

Melba's problem is to decide which of these propositions is the correct one. If she were in possession of prior probabilities and a utility function, she could maximize her expected utility. To do so, however, she would have to specify her preferences for these questions. But Melba is under obligation only to decide which question was asked, not to evaluate her personal interests in food, sports, or matrimony.

Even though she may have difficulty specifying her preferences, it should be evident to Melba that (M-3) is a much more important question than is either (M-1) or (M-2). (M-3) involves life-altering possibilities, while (M-1) and (M-2) involve only short-term prospects. (M-3) may require days or weeks for its evaluation, while the other questions may be answered with little or no deliberation. In other words, these questions possess vastly different informational value. If Melba were to reject (M-1) as the message being sent, she would have eliminated an unimportant possibility and would have narrowed her scope of concern to more weighty matters. Consequently, she would conserve a considerable amount of informational value. Rejecting (M-3), however, conserves very little informational value, since the most weighty possibility would then be dismissed, and Melba would be left to choose between questions regarding relatively inconsequential matters. Melba would thus be reluctant to reject (M-3) without significant evidence that it was *not* the question that was asked.

From the point of view of rejecting, rather than accepting, propositions, Melba has the following Boolean algebra of decisions available to her:

$$\mathcal{F} = \begin{cases} \text{reject none of the propositions,} \\ \text{reject M-1 only,} \\ \text{reject M-2 only,} \\ \text{reject M-3 only,} \\ \text{reject both M-1 and M-2,} \\ \text{reject both M-1 and M-3,} \\ \text{reject both M-2 and M-3,} \\ \text{reject all of the propositions.} \end{cases}$$

Each of these decisions will entail loss of a certain informational value as a result of its rejection. This informational value is independent of the likelihood of correctness. Melba has no strict obligation to make a unique best choice if she does not possess sufficient information to do so reliably. She can choose to defer making a choice, she can choose to reject those propositions that are of little importance to her and focus on the propositions of high informational value, or she is still free to attempt to make a "best" choice for the true proposition.

3.3   Epistemic games

Suppose a decision maker X has at its disposal a set U of propositions that are under consideration and a knowledge corpus K such that all of the elements of U are consistent with K, and that one and only one element of U is true. Note first that if X were content with the present state of affairs, there would be no incentive to consider altering its knowledge corpus, and X would therefore be unmotivated even to formulate a decision problem. Let us assume that X has a need and desire to improve its knowledge. If X were a truth-seeker as described by James, it might consider choosing the element U that maximizes the expected value of its truth-specific utility. But if X is a Jamesian error-avoider, it would not be so bold and might consider the more conservative approach of refining U by eliminating from consideration those propositions that become unattractive when held up to the light of information.

The non-rejected propositions would then all be considered reasonable possibilities to pursue.

There are two relevant desiderata to be employed by a Jamesian error-avoider when evaluating propositions for inclusion into one's knowledge corpus: namely, the desire to obtain new information and the desire to avoid error. These two criteria, although not in direct conflict, are also not in direct alignment. Consequently, we must find some way to formulate a tradeoff between these two desiderata. Our approach will be a direct application of conventional von Neumann–Morgenstern game theory – we will seek a solution that maximizes expected utility. This may seem to be a somewhat incongruous way to tackle our problem, given our desire to circumvent the superlative (substantive) in favor of the comparative (ameliorative). Objections to the use of the superlative paradigm would be well justified if we were, for example, to use it to minimize risk, as would be done under a conventional Bayesian approach. But I do not propose to do anything remotely like conventional Bayes decision theory or Neyman–Pearson theory or maximum likelihood theory or any of the other well-known solution techniques that are the workhorses of the superlative paradigm. As we shall see, that paradigm simply identifies the largest set of propositions that are justified under a comparative paradigm, with no requirement that any of them be justified according to substantive rationality.

My approach here involves the simultaneous playing of two games. The first game, which we may term the *information-conservation game*, involves X's desire to refine or narrow its choices in the interest of focusing attention on those propositions that are deemed to be the most informative. This refinement occurs independently of any error associated with the propositions. The second game, termed the *error-avoidance game*, involves X's desire to avoid focusing on propositions that are likely to be in error. X may play both games simultaneously by forming a utility function that is a weighted average of the two utilities – information and error-avoidance. The relative weighting given each game is a measure of X's concern for each.

3.3.1 The information-conservation game

In the light of abductive considerations only, X may view the worth of information without regard for error and may define a utility function to quantify informational value. A key assumption in this development is that it is possible to define a total ordering with regard to the informational value of the propositions under consideration. This leads to a very simple game between a decision maker X and Nature. Nature plays by selecting the true $u \in U$, and X plays by rejecting a set, $A \in \mathcal{F}$, where $\mathcal{F}$ is a Boolean algebra of proposition sets over U. With this very peculiar game, X does not take the possibility of error into account. Rather, X's criterion for making a selection is based on abductive considerations only.

If a proposition is of low consequence to X – that is, if it is very uninformative – it should be given a high informational value of rejection. The more propositions X

rejects, the more it can concentrate its attention on the ones that are retained. To retain only the singleton proposition with the lowest (assuming it is unique) informational value of rejection would permit X to contemplate the value associated with retaining that and only that proposition. But X can focus its considerations even more than that. To reject the entire set of propositions would provide the ultimate amount of information, for then X would be in a state of contradiction, since X's knowledge corpus would contain both the belief that one and only one of the members of U is true as well as the belief that none of them is true. This situation is possible since X, as it plays this game, is not concerned with error – only with informational value.

To define the utility of informational value of rejection, it is reasonable that the following properties should hold. Let P_R be a utility function that quantifies the informational value of rejection.

1. Since X chooses sets, rather than points, P_R should be a mapping over the Boolean algebra, $\mathcal{F}$.
2. Measures of informational value should be non-negative and finite. Consequently, it is reasonable to suppose that there exists a unit of informational value; that is, the informational-value utility should range over the unit interval.
3. Rejecting none of the propositions is of no informational value, thus $P_R(\emptyset) = 0$. Also, rejecting all propositions is of maximal informational value, thus $P_R(U) = 1$.
4. The incremental value of rejecting any set of propositions that does not intersect any previously rejected set should be invariant to whatever else has been rejected.
5. The informational value of rejecting disjoint propositions sets must be additive. That is, if $A_1 \in U$ and $A_2 \in U$ and $A_1 \cap A_2 = \emptyset$, then rejecting both A_1 and A_2 is equivalent to rejecting $A_1 \cup A_2$; that is, $P_R(A_1 \cup A_2) = P_R(A_1) + P_R(A_2)$.

This mathematical structure implies that the utility function P_R is actually a probability measure. P_R does not, however, possess any of the traditional interpretations of probability. It is not a means of quantifying such attributes as belief, propensity, frequency, etc. Rather, it is used here as a means of quantifying informational attributes. To emphasize this point, let us refer to P_R as **rejectability**, rather than probability.

Informational value is a resource. It is consumed by retaining propositions and it is conserved by rejecting them. Suppose, for some reason, we deem[3] $\{u_1\}$ to be unacceptable and reject it. By so doing, we conserve $P_R(\{u_1\})$ worth of informational value. If $P_R(\{u_1\}) \approx 1$, then we would be quite willing to discard it in the interest of conserving informational value. In effect, by rejecting u_1 we bank $P_R(\{u_1\})$ worth of informational value – it will never be consumed. If, on the other hand, $P_R(\{u_1\})$ were small, say $P_R(\{u_1\}) \approx 0$, then we would consume very little informational value by retaining it and there would be little incentive to reject $\{u_1\}$ on that basis. If we reject $\{u_2\}$ that is distinct from $\{u_1\}$ as well as reject $\{u_1\}$, the net change in informational value that accrues is the same as if we were to have simultaneously rejected $\{u_1, u_2\}$.

---

[3] Singleton sets are identified by the notation $\{\cdot\}$. Thus, whereas $u \in U$, we have $\{u\} \subset U$ and $\{u\} \in \mathcal{F}$.

3.3.2   The error-avoidance game

According to X's lights, one and only one of the members of U is true. If X is to avoid error, then it can do so only if it does not reject a set that contains the true proposition. If $u \in U$ is true, then $A \in \mathcal{F}$ is true if $u \in A$. We may express this situation mathematically by defining the so-called **indicator function**,

$$I_A(u) = \begin{cases} 1, & u \in A, \\ 0, & u \notin A. \end{cases} \tag{3.1}$$

I_A is the utility of retaining A when u is true. We may form a simple *error-avoidance game* for X playing against Nature by, as with the information-conservation game, taking U as the "proposition space" for both Nature and X. Nature chooses which proposition u is true and X chooses which set A of propositions not to reject. In consequence of X's choices, error is introduced or it is avoided. With this setup, $I_A(u)$ becomes the **error-avoidance utility**. It is the utility associated with not rejecting A if u is error-free. Note that this utility function is very even-handed; it is concerned only with the truth of the proposition. It does not count one proposition (if true) as more important, in any way, than another proposition (if true). Instead, the matter's relative importance devolves on the information-determining utility function. The error-avoidance utility function is concerned only with avoiding error and not with accounting for the seriousness of one error relative to another.

For the special case of A being a singleton set, say $A = \{u^*\}$, then $I_{\{u^*\}}(u)$ is consistent with the stance of an intransigent truth-seeker who insists on the truth and nothing but the truth; but if error-avoidance is the aim, we must consider letting A be any member of the Boolean algebra, $\mathcal{F}$.

3.3.3   Levi's epistemic game

Suppose X were a utility maximizer. If X were to play the information-conservation game alone, X would want to maximize informational value of rejection and would do this by rejecting all propositions; X would reject U. If, however, X were to play the error-avoidance game alone, X would select the only proposition guaranteed to be completely error-free; namely, X would select the entire set of propositions, U. But suppose X were to play both games simultaneously and use, as its utility, a weighted average of the two utility functions. Let us call this game *Levi's epistemic game*. Without loss of generality, we may normalize these weights to sum to unity. The resulting utility function would then be

$$\phi(A, u) = \alpha I_A(u) + (1 - \alpha)(1 - P_R(A)), \tag{3.2}$$

where $0 \leq \alpha \leq 1$. Another way to think of this situation is to perform a compound experiment by first tossing a coin with probability of heads equal to α, to play the error-avoidance game if heads lands up, and to play the information-conservation game if tails lands up. The average utility of this compound game is given by (3.2).

$\phi(A, u)$ is the **epistemic utility function** (Levi, 1980) and measures the epistemic value of *not* rejecting A when u is true. If $u \notin A$, then $\phi(A, u) = (1 - \alpha)(1 - P_R(A))$ since $I_A(u) = 0$. But if $u \in A$, then $\phi(A, u) = \alpha + (1 - \alpha)(1 - P_R(A))$, since $I_A(u) = 1$.

Since positive affine transformations, that is, transformations that consist of multiplying the original function by a positive scalar and adding an arbitrary constant, serve only to change the origin and scale of a utility function, they do not affect the preference ordering, and we may simplify the epistemic utility function by applying a transformation of the form

$$\varphi(A, u) = a_1 \phi(A, u) + a_2,$$

where $a_1 > 0$. With $a_1 = \frac{1}{\alpha}$ and $a_2 = -\frac{1-\alpha}{\alpha}$, we obtain

$$\varphi(A, u) = \frac{1}{\alpha}\phi(A, u) - \frac{1-\alpha}{\alpha}$$
$$= I_A(u) - q P_R(A), \tag{3.3}$$

where

$$q = \frac{1-\alpha}{\alpha}. \tag{3.4}$$

Since ϕ and φ are equivalent utility functions, let us also refer to $\varphi(A, u)$ as the epistemic utility function.

The epistemic utility function is a function of two variables, u and A, and represents the utility associated with Nature choosing proposition u as true, and of X choosing to reject all propositions except the elements of A. Unfortunately, since X does not know which element of U is true, X cannot evaluate the epistemic utility function. If X possesses a probability to characterize the truth support associated with the elements of $\mathcal{F}$, however, it can compute the expected value of the epistemic utility function with respect to this probability. Given a truth-supporting probability measure, termed a **credal probability** (Levi, 1980), P_S, the **expected epistemic utility** is

$$\pi(A) = \int_U \varphi(A, u) P_S(du)$$
$$= \int_U [I_A(u) - q P_R(A)] P_S(du)$$
$$= P_S(A) - q P_R(A). \tag{3.5}$$

For finite U and $\mathcal{F}$ the power set,[4] every singleton proposition is an element of $\mathcal{F}$. We may compute, for any set $A \in \mathcal{F}$, expressions for $P_S(A)$ and $P_R(A)$ in terms of mass functions

$$p_S(u) = P_S(\{u\}),$$
$$p_R(u) = P_R(\{u\}),$$

such that

$$P_S(A) = \sum_{u \in A} p_S(u),$$
$$P_R(A) = \sum_{u \in A} p_R(u).$$

If U is the set of real numbers and $\mathcal{F}$ is the Borel field, we may express credal probability and the informational value of rejection as

$$P_S(A) = \int_A p_S(u)du,$$
$$P_R(A) = \int_A p_R(u)du$$

for $A \in \mathcal{F}$. The functions p_S and p_R are Radon–Nikodym derivatives of P_S and P_R with respect to Lebesgue measure, respectively, and may be interpreted as density functions. Thus (3.5) may be written as

$$\pi(A) = \sum_{u \in A}[p_S(u) - qp_R(u)] \tag{3.6}$$

when U is a finite set and as

$$\pi(A) = \int_A [p_S(u) - qp_R(u)]du \tag{3.7}$$

when U is a continuum.

Expected epistemic utility permits us to express the dichotomy between avoiding error and acquiring information. For any set A, X would desire to retain A in the interest of avoiding error and would desire to reject A in the interest of acquiring informational value of rejection. The expected epistemic utility of A is then the difference between the belief that A is error-free and q times the informational value that would accrue if A were rejected. This difference is maximized when A is the set of all propositions for which truth support is at least as great as the informational value of rejection, that is, for the set

$$\Sigma_q = \arg\max_{A \in \mathcal{F}} \pi(A) = \{u \in U : p_S(u) \geq qp_R(u)\}, \tag{3.8}$$

---

[4] See Appendix C for a definition.

which holds for both the finite and continuum cases. The set Σ_q is the set of all propositions X views as possessing sufficient truth support relative to their informational value to risk entertaining as serious possibilities for inclusion into its knowledge corpus. All elements of $U \setminus \Sigma_q$, the complement of Σ_q, are such that they are either not likely to be true or, even if likely to be true, are not sufficiently valuable informationally to take the risk of choosing erroneously.

We may view P_S as the truth-support utility and P_R as the rejection-support utility. These two attributes generate a dichotomy: we are interested in truth, but we are also interested in informational value. Thus, X would actually render a decision based on a weighted difference of these two utilities – X evaluates a dichotomy.

The parameter q is the **index of boldness** (Levi, 1980), and characterizes the degree to which the decision maker is willing to reject informationally valuable propositions (that is, propositions with low p_R values) in the interest of avoiding error. To ensure that rejecting A erroneously cannot have higher epistemic utility than rejecting A' correctly when $P_R(A')$ is sufficiently lower than $P_R(A)$, we require that $\alpha \geq \frac{1}{2}$, which means that we require that $q \leq 1$. When $q = 1$, the decision maker rejects as many of the propositions as possible, thereby reducing the size of the set of seriously possible propositions to a minimum. As q decreases, the size of this set increases, and X becomes increasingly cautious in its propensity to reject propositions. Thus, we may equivalently view q as an **index of caution**. As boldness increases (q becomes larger), caution decreases.

Figure 3.1 illustrates the structure of this decision rule. The two utility functions, P_S and P_R, are mappings from the proposition space, U, onto a **utility space**, $\mathbf{p}(U)$, with elements consisting of (p_S, p_R) pairs for each $u \in U$, that is,

$$\mathbf{p}(u) = (p_S(u), \, p_R(u)).$$

Thus, for every element $u \in U$ there is a corresponding utility pair in the range space, $\mathbf{p}(U)$. The decision procedure generates a comparison set consisting of the region of utility space, $\mathbf{C}_q$, where truth-supporting value exceeds boldness times the informational value of rejection:

$$\mathbf{C}_q = \{(x, y) \in \mathbf{p}(U): x \geq qy\}.$$

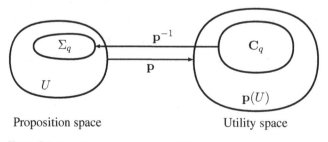

Proposition space   Utility space

Figure 3.1: Levi's rule of epistemic utility.

The inverse image of $\mathbf{C}_q$ under the utility mapping $\mathbf{p}$ composes the decision set:

$$\Sigma_q = \mathbf{p}^{-1}(\mathbf{C}_q) = \{u \in U: \mathbf{p}(u) \in \mathbf{C}_q\}.$$

As noted before, epistemic utility theory is based on von Neumann–Morgenstern game theory, and the corresponding solution is one that maximizes expected utility. This maximization, however, does not admit the usual interpretation of yielding a best solution. Rather, the maximization procedure serves only to provide the largest member of the collection of sets of propositions for which the truth support equals or exceeds the informational value gained by rejection. The elements of this maximal set, however, do not themselves necessarily possess any specific optimizing properties (they may, but it is not required). Thus they are not von Neumann–Morgenstern solutions in any sense. They are, however, intrinsically rational solutions, in that the gains (truth support) equal or exceed the losses (informational value of rejection). They are satisficing solutions. In this sense, they satisfy the adequacy desideratum, A-1, of Section 1.3.

Example 3.3 Continuing with Example 3.2, suppose Melba has a unit of informational value of rejection she wishes to apportion among the hypotheses M-1, M-2, and M-3. One possible assignment would be

$$p_R(\text{M-1}) = 0.60,$$
$$p_R(\text{M-2}) = 0.35, \tag{3.9}$$
$$p_R(\text{M-3}) = 0.05.$$

According to this assignment, Melba ascribes little informational value to the rejection of M-3, because that is an important question to be asked. She rates M-2 as being roughly twice as informationally valuable as M-1, presumably because making a decision about sports is more interesting or important to her than making a decision about cheese. We reiterate that these assignments are made independently from any beliefs as to which of the three was the question that was actually asked – the informational value is only a measure of the importance of that question to Melba. Since M-2 and M-1 are comparatively of less importance to Melba, she ascribes larger informational value to the rejection of these propositions. By invoking (3.9), Melba is able to determine the informational value of rejecting any element of the Boolean algebra $\mathcal{F}$. For example, the informational value of rejecting the disjuncts M-1 and M-3 is 0.65.

As for the specification of the credal probability, p_S, suppose Melba loves cheese, so it would be reasonable for Milo to offer her some. Also, she likes to attend ball games, and it would be reasonable for him to invite her to one. Moreover, since the two are increasingly romantically inclined, it would be reasonable for him to propose to her. If Melba has no specific information regarding the likelihood of which of the three propositions was actually sent, however, she would be in a state of belief neutrality and would be justified in adopting a least informative distribution; namely, the uniform mass function.[5] Thus, she could make the assignment

$$p_S(\text{M-}i) = \frac{1}{3}, \quad i = 1, 2, 3. \tag{3.10}$$

Melba must also make an assessment of her boldness. If she wishes to reject as many propositions as possible, she would select $q = 1$. So doing, using (3.9) and (3.10), requires that Melba reject M-1 and M-2, leaving M-3 as the only unrejected proposition. If Melba were somewhat tentative, she might set $q = 0.8$, in which case she would reject only M-1, allowing considerations of ball games and matrimony to remain. In this case, the probability of either of them being the issue raised by Milo would exceed the informational value of rejecting them as important to Melba.

Once Melba has invoked expected epistemic utility, she has used that consideration epistemic utility to the maximum extent possible in order to eliminate possibilities. If only one proposition survives the application of this rule, then it may be regarded as true. It is not generally the case, however, that even maximum boldness will result in a unique surviving hypothesis. Far from being a weakness of this approach, however, this characteristic of epistemic utility theory is a considerable strength. If there is not sufficient evidence to warrant the rejection of all but one proposition, then that situation will not occur. *There is no compulsion to demand the truth and only the truth.*

It is tempting to further analyze the surviving propositions in an attempt to determine which one is true. Melba might rightly argue: "If I am to act, I cannot very well act on a set of possibilities – surely I cannot be expected respond simultaneously to both M-2 and M-3!" But in the present context, it is important to realize that arrival at one and only one possibility is not the intent of epistemic utility. All of the surviving propositions enjoy equal status in the sense that they are all to be regarded as serious possibilities, based on their likelihood of being error-free *and* their importance to the decision maker. The available evidence has taken the decision problem as far as it can without additional information being applied. The net benefit of epistemic utility theory for Melba is to refine her beliefs by eliminating some of the propositions. The theory does not tell her how to act, nor is it intended to do so.

3.4   Praxeic utility

Praxeology is the branch of knowledge that deals with practical activity and human conduct. It is the science of efficient action (Kotarbiński, 1965). Whereas epistemology refers predominantly to the cognitive domain and concerns itself with the issue of "what to believe," praxeology lies in the practical domain and concerns itself with the issue of "how to act." Thus, whereas an epistemologist takes a set of propositions

---

[5] Melba could also refuse to adopt a numerically definite credal probability, but a discussion of that possibility will be deferred until Chapter 5.

under consideration, a praxeologist takes a set of possible actions, or *options*, under consideration. Our approach is to adapt epistemic utility theory to praxeic issues. Our interest in this theory is motivated by its striking compatibility with the comparative paradigm, since it permits costs and benefits to be characterized via probability measures that can be evaluated by the expected epistemic utility test (see Appendix D).

Although freedom from error and informational value are natural semantic notions for cognitive decision making, they are not always natural for practical decision making. To apply the ideas of epistemic utility theory to practical decision making, we must formulate praxeological analogs to the epistemological notions of avoidance of error and informational value.

Just as an epistemologically motivated decision maker has a goal of improving its state of knowledge by evaluating propositions, a praxeologically oriented decision maker has a goal of achieving some valuable objective by implementing options. Thus, a natural analog for the epistemological concept of *veracity* is the praxeological notion of *success*, in the sense of achieving some fundamental goal of taking action.

To formulate a praxeological analog for the epistemic notion of informational value, we observe that, just as the management of a finite amount of relevant information is important when inquiring after truth in the epistemological context, taking effective action requires the management of a finite amount of resource, such as time, money, materials, energy, safety, or other assets, in the praxeological context. Thus, an apt praxeological analog to the *informational value* of rejection is the *conservational value* of rejection. We thereby may rephrase Popper's injunction to become: *we want more than success – what we look for is efficient success.* Thus, we change the context of the decision problem *from one of acquiring information while avoiding error to one of conserving resources while avoiding failure.* With the context shift from the epistemological issue of belief to the praxeological issue of action, let us refer to the resulting utility function as **praxeic utility** rather than epistemic utility.

As with the epistemic game, let us refer to the degree of resource consumption as **rejectability** and require it to be expressed in terms of a function that conforms with the axioms of probability. Thus, for a finite (or continuous) option space, U, rejectability is expressed in terms of a mass function (or density function if U is a continuum), p_R such that $p_R(u) \geq 0$ for all $u \in U$ and $\sum_{u \in U} p_R(u) = 1$ (or $\int_U p_R(u)du = 1$). We will term p_R the **rejectability mass function**. Inefficient options (those with high resource consumption) should be highly rejectable; that is, if considerations of success are ignored, one should be prone to reject options that result in large costs, such as high energy consumption, exposure to hazard, etc. Normalizing p_R to be a rejectability mass function (density function) insures that the decision maker will have a unit of resource consumption to apportion among the elements of U. We may view p_R as the inutility of consuming resources. If $u \in U$ is rejected, then the decision maker conserves $p_R(u)$ worth of its unit of resources. We will often assume that $p_R(u) > 0$ for all $u \in U$ (a condition of there being no completely cost- or risk-free options).

Failure is avoided if successful options are not rejected, but efficiency, as well as success, must also be considered. Our approach to evaluating candidate sets of options for retention is to define the utility of not rejecting them in the interest of both success and resource conservation and to retain the set that maximizes this utility. Suppose that implementing $u \in U$ would lead to success, and let $A \subset U$ be a set we are considering for retention. As with epistemic utility, the utility of not rejecting A in the interest of avoiding failure is the indicator function, (3.1). We may define the praxeic utility of not rejecting A when u is successful as the convex combination of the utility of avoiding failure and the utility of conserving resources:

$$\phi(A, u) = \alpha I_A(u) + (1 - \alpha)\left(1 - \sum_{v \in A} p_R(v)\right),$$
(3.11)

where $0 \leq \alpha \leq 1$ and is chosen to reflect the decision maker's personal weighting of these two desiderata. Setting $\alpha = \frac{1}{2}$ means equal concern for avoiding failure and conserving resources.

If u is successful, then it is clear that praxeic utility is maximized when $A = \{u\}$ (the singleton set). Unfortunately, from the standpoint of resource conservation, we cannot say which u will lead to success (or that only one u will do so), so we cannot simply reject $U \setminus \{u\}$, the complement of $\{u\}$. We may, however, possess information regarding the degree of success support possessed by each u. Let p_S be a mass function (density function) that evaluates each option with respect to the degree to which it accomplishes the objective of the decision problem, independently of how much resource is consumed by implementing it. Let us refer to the degree of success support as **selectability**, and let us term p_S the **selectability mass function (density function)**. We may then calculate average praxeic utility for any set $A \subset U$ by weighting the utility by the degree of success support associated with each u and summing (integrating) over all $u \in U$. The **expected praxeic utility** is then, after making the same positive affine transformation as before,

$$\pi(A) = \sum_{u \in A} [p_S(u) - q p_R(u)],$$
(3.12)

for finite U and

$$\pi(A) = \int_A [p_S(u) - q p_R(u)]\, du,$$
(3.13)

for U a continuum, where q is given by (3.4).

We may obtain the largest set of options for which the selectability is greater than or equal to q times the rejectability by choosing the set that maximizes expected praxeic utility, resulting in the **satisficing set**

$$\Sigma_q = \arg\max_{A \subset U} \pi(A) = \{u \in U : p_S(u) \geq q p_R(u)\}.$$
(3.14)

Let us refer to (3.14) as the **praxeic likelihood ratio test** (PLRT). Σ_q is the set of all options for which the benefits (selectability) outweigh the costs (rejectability), as scaled by q. The parameter q is the **index of boldness** of the decision problem and parameterizes the degree to which the decision maker is willing to risk rejecting possibly successful options in the interest of conserving resources. Equivalently, we may also refer to q as the **index of caution**, since it parameterizes the degree to which the decision maker is willing to accommodate increased costs to achieve success.

Nominally, $q = 1$, which attributes equal weight to success and resource conservation interests. Setting $q > 1$ attributes more weight to resource conservation than to success.

Theorem 3.1

$q \leq 1 \Rightarrow \Sigma_q \neq \emptyset.$

PROOF

If $\Sigma_q = \emptyset$, then $p_S(u) < q p_R(u) \; \forall u \in U$, and hence
$1 = \sum_{u \in U} p_S(u) < q \sum_{u \in U} p_R(u) = q$, a contradiction.   □

When defining the selectability and rejectability mass (or density) functions, the decision maker must provide operational definitions of what is selectable about the options and what is rejectable. Typically, the attributes of an option that contribute to the fundamental goal of the decision problem would be associated with selectability, and those attributes that inhibit or limit activity would be associated with rejectability. However, there generally will not be a unique way to frame such decision problems. For example, suppose my decision problem is to choose a route home from work with the constraints that I wish to arrive home in a timely manner and I wish to avoid driving through dangerous parts of town. One way to frame this problem is to specify safety as a selectability attribute and driving time as a rejectability attribute. An alternative is to specify danger as a rejectability attribute and the reciprocal of driving time as a selectability attribute.

Regardless of the way the problem is framed, however, it is essential that the selectability and rejectability attributes not be restatements of the same thing. For example, it would not be appropriate to specify safety as a selectability attribute and danger as a rejectability attribute. In general, at least for single-agent decision problems, once operational definitions of success and resource consumption are specified, the selectability of an option should be specifiable without taking into consideration the consumption of resources, and the consumption of resources should be specifiable without taking success into consideration (we will need to modify this specifiability when we consider multi-agent decision problems in Chapter 6).

I have distinguished between the epistemological issues of classifying propositions in terms of their informational value and their avoidance of error and the praxeological issues of classifying options in terms of their conservational efficiency and their

Table 3.1: Epistemological and praxeological analogs

Epistemology	Praxeology
Propositions	Options
Veracity	Success
True	Instantiated
Knowledge	Resource
Probability	Selectability/Rejectability
Informational value	Conservational value

avoidance of failure. To emphasize this distinction, I have introduced new terms in order to make explicit the analogy between the cognitive and the practical domains. The epistemological and praxeological analogical relationships are characterized in Table 3.1.

With these analogies we convert the epistemological desideratum of acquiring new information while avoiding error to the praxeological desideratum of conserving resources while avoiding failure. In subsequent discussions, I will not be overly careful to distinguish between epistemic and praxeic concerns and may intermingle the two domains. To standardize notation, however, I will usually favor the praxeic terminology over the epistemic terminology. Thus, even if the context is purely epistemic, I will often substitute "selectability" for "belief" and "conservation" for "information," leaving the reader to make the proper interpretations from the context.

Example 3.4 Wendle and Ace each receive a free ticket to a ball game but there is a chance of rain. Each independently chooses either to go to the game (G), to stay home (H), or to go to the museum (M). We may view this decision problem as a game against Nature, where Nature chooses whether to rain (R) or to shine (S). The following are the possible outcomes of this game:

u_1: go to the game and it rains (R&G),
u_2: go to the game and it shines (S&G),
u_3: go to the museum and it rains (R&M),
u_4: go to the museum and it shines (S&M),
u_5: stay home and it rains (R&H),
u_6: stay home and it shines (S&H).

Each player's preferences, in descending order are

$$u_2 \succeq u_3 \succeq u_5 \succeq u_4 \succeq u_1 \succeq u_6.$$

According to the weather report, the probability of rain is β.

Wendle is an expected utility maximizer; that is, he is determined to maximize his expected enjoyment. To do so, he formulates his decision problem as a game in the traditional von Neumann–Morgenstern way by converting his preferences into a utility that is a function, ϕ, of both his and Nature's options. The values assumed by this function are as indicated in the following table:

Nature	Wendle		
	G	H	M
R	1	3	5
S	10	0	2

π	$10-9\beta$	3β	$2+3\beta$

Wendle calculates the expected utility of choosing option G as

$$\pi(G) = \beta\phi(R\&G) + (1-\beta)\phi(S\&G)$$
$$= \beta + (1-\beta)10$$
$$= 10 - 9\beta,$$

the expected utility of choosing option H as

$$\pi(H) = \beta\phi(R\&H) + (1-\beta)\phi(S\&H) = 3\beta,$$

and the expected utility of choosing option M as

$$\pi(M) = \beta\phi(R\&M) + (1-\beta)\phi(S\&M)$$
$$= 5\beta + (1-\beta)2$$
$$= 2 + 3\beta.$$

Wendle chooses the option that maximizes his expected utility. Clearly, $\pi(H) <$ $\max\{\pi(G), \pi(M)\}$, so Wendle will not choose H under any circumstances. He will choose G if $\pi(G) \geq \pi(M)$, which evaluates to $\beta \leq \frac{2}{3}$; otherwise, he will choose M.

Ace is a satisficer and proceeds according to the dictates of praxeic utility theory. He views enjoyment as a resource and seeks to conserve his enjoyment, which is governed by his preferences. Normalizing this ranking to a mass function, Ace defines his conservation-determining function (rejectability) as

$$p_R(R\&G) = 0.25,$$
$$p_R(S\&G) = 0.0,$$
$$p_R(R\&M) = 0.15,$$
$$p_R(S\&M) = 0.20,$$
$$p_R(R\&H) = 0.10,$$
$$p_R(S\&H) = 0.30.$$

(The reader may be concerned that this specification is somewhat arbitrary, and it is; there is a degree of arbitrariness in any subjective evaluation. We may gain some scant comfort in the realization that the specification of ϕ with the conventional game is also subjective and arbitrary – there is no "ultimate arbiter" in either case.)

Ace has a natural way to evaluate his selectability. Not going to the game when it shines constitutes failure, as does going to the game when it rains and staying home under any circumstances. The success support for going to the game is proportional to the probability that it will be sunny, and the success support for going to the museum is proportional to the probability that it will rain. His selectability mass function is thus

$$p_S(R\&G) = 0.0,$$
$$p_S(S\&G) = 1 - \beta,$$
$$p_S(R\&M) = \beta,$$
$$p_S(S\&M) = 0.0,$$
$$p_S(R\&H) = 0.0,$$
$$p_S(S\&H) = 0.0.$$

Ace weights selectability and rejectability equally by setting boldness equal to unity. For Ace to apply praxeic utility theory, he must be able to compare the advantages and disadvantages of his three choices. The rejectability and selectability functions as expressed above, however, are functions of both Ace's and Nature's options. We may obtain Ace's individual rejectability and selectability functions by averaging over Nature's options for each of Ace's options, yielding

$$p_R(G) = p_R(R\&G) + p_R(S\&G) = 0.25,$$
$$p_R(M) = p_R(R\&M) + p_R(S\&M) = 0.35,$$
$$p_R(H) = p_R(R\&H) + p_R(S\&H) = 0.40,$$

and

$$p_S(G) = p_S(R\&G) + p_S(S\&G) = 1 - \beta,$$
$$p_S(M) = p_S(R\&M) + p_S(S\&M) = \beta,$$
$$p_S(H) = p_S(R\&H) + p_S(S\&H) = 0.0.$$

Ace may now apply the PLRT to these marginal mass functions. Setting $q = 1$ provides equal weight for both desiderata and yields

$$\Sigma_q = \begin{cases} \{M\} & \text{if } \beta > 0.75, \\ \{G\} & \text{if } \beta < 0.35, \\ \{M, G\} & \text{if } 0.35 \leq \beta \leq 0.75. \end{cases}$$

If it is very likely to rain, then Ace's only satisficing option is M, and if it is very unlikely to rain, his only satisficing option is G. In the range of intermediate probability of rain, however, both M and G are satisficing options for Ace. But Ace must choose – he cannot go to both. Praxeic utility has narrowed the choice as much as it can but, to refine the satisficing set to a singleton, Ace must appeal to additional criteria. There are many ways to do this. For example, if Ace were to choose the satisficing option with maximum selectability, he would choose G if $\beta \leq 0.5$ and M otherwise. If he were to choose the satisficing option with minimum rejectability, he would choose G. If he were to choose the satisficing option with the greatest difference between selectability and rejectability, he would choose G if $p_S(G) - p_R(G) \geq p_S(M) - p_R(M)$, which obtains when $\beta \leq 0.55$, and would choose M otherwise. He could also choose randomly between the two satisficing options or invoke any other scheme. The point is that, *praxeic utility is not designed always to force the decision maker to a single best option. Rather, it is designed to eliminate as many unacceptable solutions as possible.*

Now let us examine these games in terms of failure. According to the doctrine of maximizing expected utility, a failure occurs if the decision does not yield the desired result – in this case, failure is going to the game if it rains, or going to the museum if the sun shines. Wendle is certain to go to the ball game if, and only if, $\beta \leq \frac{2}{3}$, in which case $P(G) = 1$; otherwise, Wendle will go to the museum, in which case $P(M) = 1$. Since weather is not affected by Wendle's attendance at either place, P_F, the probability of failure, is

$$
P_F = \begin{cases} P(R\&G) = P(R|G)P(G) = \beta & \text{if } \beta \leq \frac{2}{3}, \\ P(S\&M) = P(S|M)P(M) = 1 - \beta & \text{if } \beta > \frac{2}{3}. \end{cases}
$$

Circumstances can thus occur where Wendle has a greater probability of failing than of succeeding. Wendle is willing to take the risk, since the reward of going to the game is considerably higher than the reward for going to the museum.

Ace, however, plays his game in a way that is designed to avoid failure, rather than from the point of view of doing the "best" thing. Let's see how this works out for him. There are three cases to consider.

- When $\beta < 0.35$, $P_F = \beta$, and Ace goes to the game with a low likelihood of failure.
- When $\beta > 0.75$, $P_F = 1 - \beta$, and Ace goes to the museum with a low likelihood of failure.
- When $0.35 \leq \beta \leq 0.75$, both G and M are good enough, according to Ace's declared priorities. Each option is evaluated on its merits and both are found to be acceptable in that the gains of implementing them outweigh the losses. Thus, Ace will not fail if he attends the museum, even if it does not rain. He will, however, fail if he goes to the game and it rains. To evaluate the probability of failure in this case requires the specification of the probability that he will go to the game, given that both the game and the museum are satisficing. Suppose he tosses a fair coin and goes to the game if it lands heads. In this case, the probability of failure will be $\beta/2$.

There are a number of significant differences between these two frameworks for decision making. Wendle forms a single utility and sets about to maximize his expected enjoyment. He has a rigid notion of success. Ace, however, forms two utilities and compares them to each other as a means of evaluating each option for the purpose of conserving as much enjoyment as he can while avoiding getting wet. He has a much more flexible notion of success. Wendle has no choice but to take the option that maximizes his expectations, while Ace affords himself some flexibility in making his final choice. Wendle's fate as an optimizer was determined the moment he decided to be an expectation maximizer, while Ace's fate as a satisficer is not completely determined until the moment of truth, when he actually makes his choice.

By keeping the concepts of avoiding failure and consuming resources separate, the decision maker is able to evaluate the relative strengths of these two attributes of each option. For example, if an option that conserves resources (low rejectability) also avoids failure (high selectability), then the option would be attractive to the decision maker under both criteria. Conversely, an option with high rejectability and low selectability would be very unattractive. Situations where selectability and rejectability are both high or both low, however, are more difficult to categorize. In Chapter 5 I discuss these four situations in detail.

3.5   Tie-breaking

Although decision making under praxeic utility is not designed to return a single decision, if action is to be taken, one and only one $u \in \Sigma_q$ must ultimately be invoked. The fact that praxeic utility theory does not provide a unique solution is not a defect of the decision-making procedure. If it is truly a comparative paradigm, praxeic utility theory should not force the superlative outcome of identifying a unique solution which, if only by default, is considered to be best.

Praxeic utility theory provides only a partial ordering of the set of options. We can only say, on the basis of the PLRT, that elements of Σ_q are preferred to elements in $U \setminus \Sigma_q$. This approach does not impose an ordering of the elements of Σ_q. But, at the moment of truth when action can no longer be deferred, some form of tie-breaking must be imposed by the decision maker. Although tie-breaking is an additional decision that must be made, it constitutes a choice between a set of alternatives where each one is known to be good enough. In this sense, the final decision is, if not necessarily easy, at least reassuring, in the sense that the decision maker cannot make a fundamentally bad choice.

One way to proceed with the design of a tie-breaker is to define an ordering of the elements of the satisficing set. There are many possible orderings, with the most obvious being the selectability and rejectability functions.

Definition 3.1

A satisficing option u_S is **most selectable** if

$$u_S = \arg\max_{u \in \Sigma_q}\{p_S(u)\}. \tag{3.15}$$

□

Definition 3.2
A satisficing option u_R is **least rejectable** if

$$u_R = \arg \min_{u \in \Sigma_q} \{p_R(u)\}. \tag{3.16}$$

□

A compromise between these two extremes is a linear combination of selectability and rejectability.

Definition 3.3
A satisficing option u^* is **maximally discriminating** if

$$u^* = \arg \max_{u \in \Sigma_q} \{p_S(u) - q p_R(u)\}. \tag{3.17}$$

□

The notions of most selectable, most rejectable, and discrimination provide partial orderings of the members of the satisficing set such that at least one element of the set is preferred to all others. Adopting such auxiliary criteria does not constitute a reversion to the optimality paradigm. The choice of tie-breaking criterion does not influence the solution methodology that produced the satisficing set and may be changed without affecting the structure of that set. Thus the decision maker can delay the application of the tie-breaker until it is necessary to take action.

Another form of partial ordering of the satisficing set is as follows:

Definition 3.4
A satisficing option u_1 is **more satisficing** than option u_2 if it is either (a) not less selectable and less rejectable than u_2 or (b) not more rejectable and more selectable than u_2; that is, either

$$p_S(u_1) \geq p_S(u_2) \quad \text{and} \quad p_R(u_1) < p_R(u_2)$$

or

$$p_S(u_1) > p_S(u_2) \quad \text{and} \quad p_R(u_1) \leq p_R(u_2).$$

□

Unfortunately, the set of pairs of options that satisfy the more-satisficing criterion may be empty. For this reason, it is not of general utility for choosing a tie-breaker.

Definition 3.5
A satisficing option is **arbitrary** if it is chosen randomly according to some probability distribution over the set of satisficing options.

□

While an arbitrary tie-breaker may provide acceptable results, such a procedure seems unnecessarily capricious in practice, since there are often auxiliary considerations that, though not part of the performance criteria, can reflect auxiliary preferences of the decision maker. In Section 4.1 I discuss a concept of equilibria that augments the basic satisficing notion and provides a way to refine the satisficing set in a natural way.

If the cardinality of Σ_q is large, computational resources may not permit identifying all members of this set, and a stopping rule must be established. One approach, in the spirit of Simon-like satisficing, is to set an aspiration level for the number of elements in Σ_q. The most simple aspiration is $\Sigma_q \neq \emptyset$, in which case the first satisficing option that is identified would be implemented. A more sophisticated criterion would be to invoke the concept of ordinal optimization (Ho, 1997), and stop when sufficiently many elements of Σ_q have been identified to ensure that, with high probability, the maximally discriminating option has been identified. Alternatively, we could employ dominance to eliminate non-Pareto efficient (with respect to selectability and rejectability) options prior to making a decision (Goodrich et al., 1998).

By way of review, a decision maker who makes choices by comparing gains and losses acts in a way that is fundamentally distinct from one who searches for a global optimum. The latter proceeds by ranking the options to determine the maximal element in a preference ordering, while the satisficer proceeds by comparing two sets of preferences and making a binary decision to either reject or not reject each individual option. Optimization requires making global rankings, while satisficing requires making local comparisons. Optimization is designed to identify a single best option; satisficing is designed to identify a set of good enough options.

The boldness parameter, q, governs the size of the satisficing set. Setting $q < 1$ ascribes more weight to achieving success than to conserving resources. Unlike the case for cognitive decisions, however, it is rational in the praxeic context for q to be greater than unity. Setting $q > 1$ simply means that conserving resources is weighted heavier than achieving success, which may be appropriate if the cost of taking action is high relative to the benefit. As q increases, the size of Σ_q decreases. Eventually, as q becomes sufficiently large, no options will be satisficing and Σ_q will be empty. Let us assume in what follows, however, that q is such that $\Sigma_q \neq \emptyset$.

3.6   Praxeology versus Bayesianism

The conventional Bayesian approach for deciding between singleton propositions $\{v\}$ and $\{u\}$ is to define a cost function

$$L(v, u) = \text{cost of choosing } v \text{ when } u \text{ is true,} \tag{3.18}$$

where we assume that $L(v, u) > L(u, u)$ for $v \neq u$ (that is, the cost of choosing incorrectly is larger than the cost of choosing correctly). This structure may be expressed more conveniently as

$$L(v, u) = a_1(v, u)[1 - I_{\{v\}}(u)] + a_2(u)I_{\{v\}}(u), \tag{3.19}$$

where u is true and v is chosen and $I_{\{v\}}(u)$ is the indicator function for the singleton set $\{v\}$. Thus, the decision maker incurs a cost of $a_2(u)$ when $u = v$ and a cost of $a_1(v, u)$ when $u \neq v$. Viewing P_S as a prior probability, we may define a Bayes rule as any element of U that minimizes the expected loss, denoted

$$\bar{L}(v) = \int_U L(v, u) P_S(du),$$

whence a Bayes decision is

$$u_{Bayes} = \arg \min_{v \in U} \bar{L}(v). \tag{3.20}$$

Thus, Bayesian analysis provides a unique (up to an equivalence class) best option that minimizes the expected loss. Note that this provides a total ordering of the elements of U. By contrast, maximizing expected praxeic utility yields the satisficing set (3.14), which provides only a partial ordering in that it differentiates between elements of Σ_q and its complement (satisficing versus not satisficing), but does not order the members of the satisficing set.

The Bayesian utility is a function of singletons (points), but praxeic utility is a function of sets. This fundamentally different structure accommodates the fact that Bayesian utility is designed to facilitate inter-option comparisons with the objective of optimizing, but praxeic utility is designed to accommodate intra-option comparisons with the objective of satisficing. Thus, the two utility functions are designed for fundamentally different purposes.

Even when restricting praxeic utility to singletons, the two utility concepts are not equivalent, in general. We may see this by noting that, for $A = \{v\}$, equating $\phi(A, u)$ (see (3.3) and $L(v, u)$ requires that

$$[a_2(u) - a_1(v, u)]I_{\{v\}}(u) + a_1(u, v) = I_{\{v\}}(u) - qp_R(v),$$

which requires that

$$a_2(u) - a_1(v, u) = 1$$
$$a_1(u, v) = qp_R(v)$$

for all $v \in U$ and all $u \in U$. This can happen only if a_1 is not a function of v, which implies that $p_R(v)$ must be constant, that is, p_R is uniform. Thus, Bayesian utility can be made to be equivalent to praxeic utility only in the special case of conservational value neutrality. For example, if U is finite with cardinality m and

$$a_1(u, v) = \frac{1}{m},$$

$$a_2(u) = \frac{m+1}{m},$$

$$q = 1,$$

$$p_R(v) = \frac{1}{m},$$

then the two utilities are the same in structure. Thus, decisions made under non-neutral conservational value must be approached from a point of view significantly different from that used to examine decisions made under conditions of value neutrality. The standard Bayesian cost structure accounts for only conditional cost, that is, where the cost is a function of both the choice that is actually made and the correct choice. The use of two utilities in praxeic utility permits the independent specification of the selectability cost (which accounts for the avoidance of failure) and the rejectability cost (which accounts for the conservation of resources regardless of correctness).

4 Equanimity

It is no paradox to say that in our most theoretical moods we may be nearest to our most practical applications.
Alfred North Whitehead
An Introduction to Mathematics (Oxford University Press, 1948)

A dutiful decision maker may not be persuaded to adopt a satisficing solution just because the gains exceed the losses. The satisficing options should also conform to a sense of fairness or equanimity. There are three additional criteria that should govern the ultimate selection of a satisficing option. First, if time and resources permit, a decision maker should neither sacrifice quality needlessly nor pay more than is necessary. Second, a decision maker should be as certain as possible that the decision really is good enough, or adequate. Third, a decision maker should not foreclose against optimality; that is, the optimal decision, should it exist, ought to be satisficing.

4.1 Equilibria

Although the satisficing set Σ_q contains all possible options that satisfy the PLRT and, in that sense, are legitimate candidates for adoption, they generally will not be equal in overall quality. Consider the following example.

Example 4.1 Lucy is in the market for a car. To keep the problem simple, assume that her set of possibilities consists of five choices, which we denote as vehicles A through E. The option space is the set $U = \{A, B, C, D, E\}$. Only three criteria are important: performance, reliability, and affordability. Suppose that Lucy is able to assign ordinal rankings to the vehicles in each of these attributes, as illustrated in Table 4.1. Vehicle B, for example, has the best performance, the median reliability, and the highest cost.

There are a number of questions that may be asked regarding the issue of how to choose from among the options presented in Example 4.1. The optimizer's question is perhaps the most direct: "What is the best deal?" To address this question, we must

Table 4.1: Ordinal rankings of vehicle attributes

Vehicle	Performance	Reliability	Affordability
A	3	1	5
B	5	3	1
C	2	4	4
D	1	5	3
E	4	2	2

Key: 5 = best; 4 = next best; 3 = median; 2 = next worst; 1 = worst

Table 4.2: Global preference and normalized gain/loss functions

Vehicle	Global preference (J)	Normalized gain (p_S)	Normalized loss (p_R)
A	9	0.133	0.067
B	9	0.267	0.333
C	10	0.200	0.133
D	9	0.200	0.200
E	8	0.200	0.267

define a preference function. Let us define this function as the equally-weighted sum of the ordinal rankings of the three attributes; that is,

$$J = \text{Performance} + \text{Reliability} + \text{Affordability},$$

where the affordability number is ordered such that the cheaper vehicle has the higher affordability. The values of this preference function are displayed in the second column of Table 4.2. Clearly, the uniquely optimal option is vehicle C, the choice that, although next worst in performance, is next best in both reliability and price. But asking for the best deal is not the only rational question one might compose. For example, suppose Lucy were to frame her question as: "Am I really going to get what I pay for?" This question does not involve making inter-comparisons among options; rather, it involves intra-comparisons of attributes for each individual vehicle. To make these intra-comparisons, a natural procedure is to separate the attributes into two categories: one to involve the attributes that represent gains to the decision maker as a result of adopting the option, and the other to involve attributes that represent losses. A natural, but not unique, categorization of this problem is to identify performance and reliability as gains, and cost as loss. To compare gains and losses, they must be represented on the same scale. This may be done by creating selectability and rejectability functions and normalizing the problem so that the decision maker has a unit of gain utility and a unit of loss utility to apportion among the options. We may do this by normalizing ordinal rankings associated with each category, yielding the last two columns of Table 4.2,

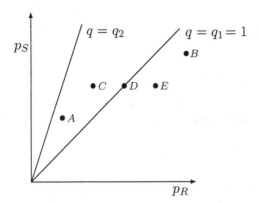

Figure 4.1: Cross-plot of selectability and rejectability.

where we have reversed the ordering on the affordability attribute to convert it to cost. These values constitute the selectability and rejectability functions, respectively. We observe that selectability exceeds rejectability for options A and C, the gain equals the loss for D, and the loss exceeds the gain for B and E.

Figure 4.1 provides a cross-plot of selectability versus rejectability as u is varied over its domain, with p_R the abscissa and p_S the ordinate. Observe that, although A has the lowest selectability, it also has the lowest rejectability, and a rational decision maker can legitimately come to the conclusion that this is satisficing, since the benefits at least outweigh the costs. Option B is at the other extreme. It has the highest performance, but is also the most expensive. In the value system of the customer, however, the benefits do not outweigh the costs, and the option is not satisficing. Option E is also easily eliminated by the cost-benefit test. Now consider options C and D. According to the PLRT, both are satisficing, and the customer would be justified in choosing either one. Choosing D, however, would cost more than C without offering increased benefit. Clearly, C should be preferred to D. It seems necessary, therefore, to classify the satisficing solutions in more detail.

This example prompts a refinement of the satisficing set to eliminate satisficing options that are dominated by other satisficing options. For every $u \in U$ let

$$B_S(u) = \{v \in U: p_R(v) < p_R(u) \quad \text{and} \quad p_S(v) \geq p_S(u)\}$$
$$B_R(u) = \{v \in U: p_R(v) \leq p_R(u) \quad \text{and} \quad p_S(v) > p_S(u)\},$$

and define the set of options that are *strictly better* than u:

$$B(u) = B_S(u) \cup B_R(u);$$

that is, $B(u)$ consists of all possible options that are either less rejectable and not less selectable than u, or are not more rejectable and more selectable than u. If $B(u) = \emptyset$, then no options can be preferred to u in both selectability and rejectability. Such a u is an equilibrium option (Stirling et al., 2002).

Definition 4.1

The set of **equilibrium options** is

$$\mathfrak{E} = \{u \in U: B(u) = \emptyset\}. \qquad \square$$

Definition 4.2

We define the set of **satisficing equilibrium** options as

$$\mathfrak{S}_q = \mathfrak{E} \cap \Sigma_q. \qquad \square$$

Theorem 4.1

If U is closed and Σ_q is not empty, then $\mathfrak{S}_q$ is non-empty.

PROOF

Suppose $\mathfrak{E} = \emptyset$. Then for every $u \in U$ either $B_S(u) \neq \emptyset$ or $B_R(u) \neq \emptyset$. Let u^* be the maximally discriminating option, that is,

$$u^* = \arg\max_{u \in U}\{p_S(u) - qp_R(u)\}. \qquad (4.1)$$

Since p_S and p_R are bounded and U is closed, $u^* \in U$. We must establish that $u^* \in \mathfrak{E}$. Suppose $B_S(u^*) \neq \emptyset$. Then there exists a $v \in U$ such that

$$qp_R(v) < qp_R(u^*) \quad \text{and} \quad p_S(v) \geq p_S(u^*).$$

But this implies that

$$p_S(v) - qp_R(v) > p_S(u^*) - qp_R(u^*),$$

which contradicts (4.1); thus $B_S(u^*) = \emptyset$. A similar argument establishes that $B_R(u^*) = \emptyset$. Consequently, $B(u^*) = \emptyset$, so $u^* \in \mathfrak{E}$ and $\mathfrak{E} \neq \emptyset$.

Since $p_S(u) - qp_R(u) \geq 0$ for all $u \in \Sigma_q$ and $\Sigma_q \neq \emptyset$, it follows that

$$p_S(u^*) - qp_R(u^*) \geq 0, \text{ so } u^* \in \Sigma_q. \text{ Thus } u^* \in \mathfrak{S}_q \text{ and } \mathfrak{S}_q \neq \emptyset. \qquad \square$$

Clearly, the maximally discriminating option is a satisficing equilibrium point. If $q = 0$, the most discriminating option is also the most selectable satisficing option, u_S, (see (3.15)). This limiting case represents a very aggressive stance to achieve the goal at the risk of excessive cost. A most selectable satisficing option may be considered for cases with large variations in p_S and small variations in p_R.

Another limiting case occurs as $q \to \infty$, resulting in the least rejectable satisficing option, u_R (see (3.16)). Adopting the least rejectable option is itself very conservative and reflects a willingness to compromise the fundamental goal in the interest of reducing cost. It may be appropriate when there are large variations in p_R relative to small variations in p_S.

As additional examples of satisficing equilibria, consider specifying a rejectability level $\rho \in (0, 1)$ and implementing the satisficing option that maximizes the selectability subject to the constraint that the rejectability does not exceed ρ. Let

$$\Sigma_q^\rho = \{u \in \Sigma_q: p_R(u) \leq \rho\}.$$

The most selectable option of rejectability ρ is then

$$u_\rho = \arg \max_{u \in \Sigma_q^\rho} p_S(u).$$

If $\Sigma_q^\rho = \emptyset$, then a most selectable option of rejectability ρ does not exist. Alternatively, one could implement the satisficing option that minimizes rejectability subject to a selectability constraint to obtain a least rejectable option of selectability α:

$$u_\alpha = \arg \min_{u \in \Sigma_q^\alpha} p_R(u),$$

provided $\Sigma_q^\alpha \neq \emptyset$, where, for $\alpha \in (0, 1)$,

$$\Sigma_q^\alpha = \{u \in \Sigma_q: p_S(u) \geq \alpha\}.$$

An important possibility emerges in the continuous case, when p_S and p_R are density functions. Suppose p_S is concave over U. That is, for $\lambda \in [0, 1]$,

$$\lambda p_S(u_1) + (1 - \lambda)p_S(u_2) \leq p_S(\lambda u_1 + (1 - \lambda)u_2) \text{ for all } u_1 \in U, u_2 \in U.$$

Also, suppose that p_R is convex over U; that is, for $\lambda \in [0, 1]$,

$$\lambda p_R(u_1) + (1 - \lambda)p_R(u_2) \geq p_R(\lambda u_1 + (1 - \lambda)u_2) \text{ for all } u_1 \in U, u_2 \in U.$$

For these classes of density functions, the following properties are important in developing a synthesis procedure.

Lemma 1

For continuous and concave selectability p_S and continuous and convex rejectability p_R, $\mathfrak{E} \subset \{u \in U: p_S^+(u)p_R^+(u) \geq 0\} \cup \{u_S, u_R\}$ where

$$p_S^+(u) = \lim_{\lambda \downarrow 0} \frac{p_S(u + \lambda) - p_S(u)}{\lambda}$$

denotes the right derivative of p_S, and similarly define P_R^+ as the right derivative of p_R.

PROOF

Since u_S and u_R represent the most discriminating controls for $q = 0$ and $q \to \infty$, respectively, by the proof of Theorem 4.1, they are in $\mathfrak{E}$. For any other $u \in \mathfrak{E}$, concavity of p_S and convexity of p_R imply that selectability and rejectability are either both non-increasing or both non-decreasing in the neighborhood of u. Hence, $p_S^+(u)p_R^+(u) \geq 0$. $\square$

If, in addition to being continuous, both p_S and p_R are differentiable with derivatives denoted $\dot{p}_S$ and $\dot{p}_R$, respectively, the equilibrium set satisfies

$$\mathfrak{E} \subset \{u \in U: \dot{p}_S(u)\dot{p}_R(u) \geq 0\} \cup \{u_S, u_R\}.$$

Moreover, for concave p_S and convex p_R the following lemma establishes a necessary and sufficient condition for determining the equilibrium set.

Lemma 2

For differentiable concave p_S and differentiable convex p_R

$$\mathfrak{E} = \{u \in U: \dot{p}_S(u)\dot{p}_R(u) \geq 0\} \cup \{u_S, u_R\}.$$

PROOF

Let $u \in \{v \in U: \dot{p}_S(v)\dot{p}_R(v) \geq 0\} \cup \{u_S, u_R\}$. If $u = u_S$ or $u = u_R$ then $u \in \mathfrak{E}$ since the most discriminating control is always in the equilibrium set. Otherwise, u must satisfy $\dot{p}_S(u)\dot{p}_R(u) \geq 0$ which implies that $\exists \lambda \geq 0$ such that $\dot{p}_S(u) - \lambda \dot{p}_R(u) = 0$. However, since the sum of two concave functions ($p_S(u)$ and $-\lambda p_R(u)$) is also concave then the u which satisfies $\dot{p}_S(u) - \lambda \dot{p}_R(u) = 0$ is a most discriminating control for $q = \lambda$. Since a most discriminating control cannot be dominated, $u \in \mathfrak{E}$. Thus,

$$\mathfrak{E} \supset \{u \in U: \dot{p}_S(u)\dot{p}_R(u) \geq 0\} \cup \{u_S, u_R\}.$$

This result, coupled with Lemma 1, establishes the desired result.   □

Theorem 4.2

Let $U \subset \mathbb{R}$. For p_S a concave density function and p_R a convex density function over U, the satisficing set Σ_q is convex. Moreover, for concave differentiable selectability density and convex differentiable rejectability density, the equilibrium set $\mathfrak{E}$ and the satisficing equilibrium set $\mathfrak{S}_q$ are convex.

PROOF

Convexity of the satisficing set is shown by establishing that, for $u_1 \in \Sigma_q$ and $u_2 \in \Sigma_q$, the point $v_\lambda = \lambda u_1 + (1 - \lambda)u_2 \in \Sigma_q$ for any $0 \leq \lambda \leq 1$. By concavity of p_S and convexity of p_R, we get that

$$p_S[\lambda u_1 + (1 - \lambda)u_2] \geq \lambda p_S(u_1) + (1 - \lambda)p_S(u_2),$$
$$-p_R[\lambda u_1 + (1 - \lambda)u_2] \geq -\lambda p_R(u_1) - (1 - \lambda)p_R(u_2),$$

hence

$$p_S[\lambda u_1 + (1 - \lambda)u_2] - q p_R[\lambda u_1 + (1 - \lambda)u_2] \geq \lambda[p_S(u_1) - q p_R(u_1)]$$
$$+ (1 - \lambda)[p_S(u_2) - q p_R(u_2)].$$

Since $u_1, u_2 \in \Sigma_q$ we know that $p_S(u_1) - qp_R(u_1) > 0$ and $p_S(u_2) - qp_R(u_2) > 0$, whence

$$p_S(v_\lambda) - qp_R(v_\lambda) > 0.$$

Thus Σ_q is convex.

To establish the convexity of $\mathfrak{E}$, first note that, by Lemma 1,

$$\mathfrak{E} \subset \{u \in U: \dot{p}_S(u)\dot{p}_R(u) \geq 0\} \cup \{u_S, u_R\}.$$

If $u_S = u_R$ then $\mathfrak{E} = \{u_S\}$. Otherwise, let $u_1, u_2 \in \mathfrak{E}$ with $u_1 < u_2$, and suppose $\dot{p}_S(u_1)$ and $\dot{p}_S(u_2)$ are of opposite sign. Since p_S is concave, this requires that $\dot{p}_S(u_1) > 0$ and $\dot{p}_S(u_2) < 0$. For u_1 and u_2 to be elements of $\mathfrak{E}$ requires that $\dot{p}_R(u_1) > 0$ and $\dot{p}_R(u_2) < 0$, which is impossible since p_R is convex. Thus the directional derivatives of p_S and p_R cannot change sign in $\mathfrak{E}$. Since p_S is concave, there can be at most one sign change in the derivative, so it can be concluded that $\dot{p}_S(v_\lambda)$ has the same sign as $\dot{p}_S(u_1)$ and $\dot{p}_S(u_2)$. A similar argument holds for $\dot{p}_R(v_\lambda)$ and, consequently, $\dot{p}_S(v_\lambda)\dot{p}_R(v_\lambda) \geq 0$. So, by Lemma 2, $v_\lambda \in \mathfrak{E}$. Finally, since the intersection of convex sets is convex, $\mathfrak{S}_q$ is convex. $\square$

This theorem means that, for concave selectability and convex rejectability defined on an interval $U = [u_{min}, u_{max}]$, the maximally satisficing set is also an interval. Moreover, $\mathfrak{E} = [\min\{u_S, u_R\}, \max\{u_S, u_R\}]$.

The following theorem establishes an equivalence between the equilibrium set and the set of most discriminating solutions. This theorem is useful because it says all elements of $\mathfrak{E}$ (not just the endpoints, u_S and u_R, are maximizing elements, whereas $\mathfrak{S}_q$ contains only satisficing *and* maximizing elements.

Theorem 4.3

Let $U \in \mathbb{R}$, let p_S be a concave selectability density, differentiable over the interior of U, and let p_R be a convex rejectability density, differentiable over the interior of U. Then, for every $u \in \mathfrak{E}$, there exists a boldness value $q \in [0, \infty)$ such that u is a most discriminating satisficing option. Furthermore, if $p_S - qp_R$ is strictly concave for every $q \in [0, \infty)$, then, for every such q, there corresponds a unique most discriminating satisficing option.

PROOF

Let $u \in \mathfrak{E}$ and define

$$u_* = \min\{u_S, u_R\},$$
$$u^* = \max\{u_S, u_R\}.$$

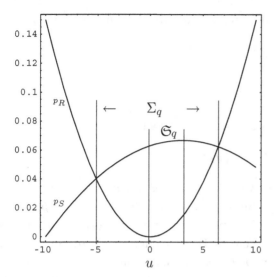

Figure 4.2: Satisficing equilibrium regions for a concave p_S and convex p_R.

For these limiting cases, it has previously been established that by assigning boldness values of $q \to \infty$ and $q = 0$, respectively, u_R and u_S are most discriminating satisficing controls. Fix $u \in (u_*, u^*)$. The function $p_S(v) - qp_R(v)$ is extremized when $\dot{p}_S(v) - b\dot{p}_R(v) = 0$. If $\dot{p}_R = 0$ then $\dot{p}_S = 0$ whence, by concavity of p_S and convexity of p_R, $v = u_A = u_L$ for any q. Otherwise, when $\dot{p}_R \neq 0$, evaluating $\dot{p}_S(v) - q\dot{p}_R(v) = 0$ at $v = u$ implies that $q_u = \frac{\dot{p}_S(u)}{\dot{p}_R(u)}$ is the boldness required to render u a most discriminating satisficing control. Furthermore, when $p_S - qp_R$ is strictly concave for all $q \in [0, \infty)$, $u = \arg\max_{v\in(u_*,u^*)}\{p_S(v) - qp_R(v)\}$ is the unique most discriminating satisficing option.   □

Figure 4.1 illustrates the satisficing and satisficing equilibrium sets for a convex differentiable p_R and concave differentiable p_S, with the selectability and rejectability as functions of u, for $q = 1$. In the figure, Σ_q consists of those u for which $p_S(u)$ exceeds $p_R(u)$. The set $\mathfrak{S}_q$ consists of those u for which no option with higher selectability exists for a given rejectability level and for which no option with lower rejectability exists for a given selectability level. Observe that for $u \in \mathfrak{S}_q$ the selectability and rejectability have slopes with the same sign (see Lemma 2); whence options in this region are in equilibrium.

4.2   Adequacy

Robert Browning's sentiment that "a man's reach should exceed his grasp" (Browning, 1855) may be good psychology, but it is bad engineering. People may be motivated by aspirations that are higher than their abilities, but artificial systems had better do the job they are designed to do. This realization has prompted engineers to rely heavily on optimization as a design principle. If a system is designed to be the very best possible, then, it is reasoned, either it will do the job or the job cannot be done. There is safety in optimality.

There is also some danger. A tunnel-vision search for the best possible solution blinds one to the existence of many other possible solutions that would be adequate. And adequacy is often the real goal, not optimality. Optimality is sometimes simply a way to insure adequacy.

For some applications, it may be possible or desirable to define minimum standards, as is done in the case of extrinsic satisficing (see Section 1.3.1). For simple problems, these minimum standards may be imposed *a priori*, and any options that do not meet them may be eliminated without further ado, leaving only the securely satisficing options. For more complex situations, however, minimum standards may be difficult, if not impossible, to ascribe *a priori*, but they may be ascertained *a posteriori*, that is, after the satisficing set has been defined. Minimum standards imposed at this point may serve to reduce the size of the satisficing set. To be so imposed will first require the standards to be translated into selectability and/or rejectability thresholds. Let p_S^* and p_R^* denote selectability and rejectability thresholds such that, if $p_S(u) < p_S^*$ or $p_R(u) > p_R^*$, then u must be rejected, even if $p_S(u) > p_R(u)$. Then the restricted satisficing set is of the form

$$\Sigma_q^r = \{u \in U: p_S(u) > p_R(u),\ p_S(u) > p_S^*,\ p_R(u) < p_R^*\}.$$

If the absolute standards are sufficiently high, then Σ_q^r may be empty, in which case it will be necessary to reduce boldness enough to accommodate the minimum standard. In Example 4.1, Lucy may *a posteriori* impose an arbitrary threshold on performance and reliability such that $p_S(A) < p_S^*$. In this case, option A must be rejected, even though it is satisficing. If she were to increase p_S^* sufficiently, she would be forced also to reject option C as well, leaving no options in the restricted satisficing set. The only way out of this predicament is for her to lower her standards of what is satisficing, that is, to reduce boldness. By reducing q sufficiently, she will eventually conclude that option B is satisficing relative to this new standard. However, B must also conform to Lucy's *a posteriori* standards regarding cost; if not, then she must conclude that there is no vehicle on the market that meets her expectations.

Satisficing as implemented under intrinsic rationality is a principled way to make dichotomous tradeoffs. Tradeoffs are the most fundamental kinds of decisions, and such activity cannot be eliminated from even the seemingly most sophisticated forms of decision making such as optimization, although they may manifest themselves in indirect and even *ad hoc* ways. For example, every control engineer well knows that specifying the performance criterion and solving for the optimum is only the first step in the subjective exercise of controller design. Consider, for example, the elementary control problem of designing an optimal linear quadratic regulator, where the goal is to maintain the state of a linear dynamical system within acceptable deviations from a desired set point while keeping the cost of control at an acceptable level. The tradeoff between performance and cost is accomplished by adjusting the relative weights of these two desiderata. Each weighting ratio leads to a different optimal control policy, and the task for the designer is to tune the weights iteratively to achieve an acceptable balance. In reality, there is no universally optimal solution – the "optimal" solution technique is nothing more than a convenient and systematic design procedure to achieve a solution that can be defended, at the end of the day, as being "good enough."

Thus, even though satisficing is based on a different notion of rationality than conventional decision making, in practice it may not be as different as it might first appear. Tradeoffs are to be made in both cases, the main difference is when and how they are made. With substantive rationality based approaches, the tradeoffs are made by adjusting the numerical weights of the utility functions as a result of (possibly *ad hoc*) evaluations of the solutions. When the designer is satisfied that the performance justifies the cost, the design is accepted, and the solution is declared to be "optimal." With intrinsic rationality, tradeoffs are made by directly comparing the benefits against the costs, and all options that satisfy that well-defined condition are admissible. Is one approach more "rational" than another? Ultimately, that is a subjective judgment itself.

4.3   Consistency

The major difference between satisficing and optimizing is that satisficing requires the formation of dichotomous tradeoffs, while optimization requires some form of minimization or maximization that lacks intrinsic dichotomies. Both approaches, however, may be governed by exactly the same criteria. If they are, then it would be reasonable to expect that the optimal decision would also be a satisficing decision. In fact, it would be difficult to have confidence in any satisficing methodology that did not generate consistency of this type.

Every decision problem involves polarization. When considering any option, the decision maker may evaluate it both in terms of the degree to which it satisfies the demands of the problem and in terms of how much it costs to implement it. The tradeoff between these two desiderata is the essence of decision making, regardless of the mechanism used to form the decision. Praxeic utility theory keeps the two desiderata separate, but optimization combines them into a single performance index. The way this combining is done is somewhat arbitrary; the essential thing to ensure is that due care be given to the relative weighting of the two desiderata that are to be combined.

It is simple to show that satisficing does not foreclose against optimization. Assuming that $\Sigma_q \neq \emptyset$, we may define the utility function

$$J(u) = p_S(u) - qp_R(u).$$

Let $u^* = \arg\max_{u \in U} J(u)$. But $J(u) \geq 0 \ \forall u \in \Sigma_q$ and, since $\Sigma_q \neq \emptyset$, $J(u^*) \geq \max_{u \in \Sigma_q} J(u) \geq 0$, which implies $u^* \in \Sigma_q$. Thus, there exists an optimal solution corresponding to exactly the same criterion as is used to define the set of satisficing solutions.

The following discussion pertains to the analysis of a discrete-time dynamic system and establishes the consistency of satisficing solutions for the quadratic regulator. The remainder of this chapter may be skipped without loss of continuity in the book's explication of praxeic utility theory.

Example 4.2 Consider a discrete-time system governed by the linear difference equation

$$x_{k+1} = Ax_k + Bu_k, \quad k = 0, 1, \ldots, t - 1.$$

x_k is an n-dimensional column vector called the state vector of the system, A is an $n \times n$ matrix, u_k is the control variable, and B is an $n \times 1$ matrix. The control problem is to choose the sequence $\{u_0, u_1, \ldots, u_{t-1}\}$ to drive the state from arbitrary initial conditions at $k = 0$ toward the origin at $k = t$ in a way that conserves energy consumed by the control variable.

We see that there are two separate desiderata: the error at the terminal time is given by[1]

$$e_1 = x_t^T x_t,$$

and the total energy consumed by the control variable is

$$e_2 = \sum_{k=0}^{t-1} u_k^2.$$

The approach taken by conventional optimal control theory is to form a weighted sum of these two desiderata, namely,

$$J(u_0, u_1, \ldots, u_{t-1}) = e_1 + Re_2$$
$$= x_t^T x_t + \sum_{k=0}^{t-1} Ru_k^2,$$

and to solve for the sequence $\{u_0, u_1, \ldots, u_{t-1}\}$ that minimizes J. The solution to this problem is one of the elegant results of classical optimal control theory and provides a feedback solution of the form

$$u_k = -K_k x_k,$$

where K_k is the Kalman gain matrix (see Lewis (1986)) given by

$$K_k = [B^T P_{k+1} B + R]^{-1} B^T P_{k+1} A$$

---

[1] The superscript T denotes the matrix transpose.

with

$$P_k = A^T P_{k+1}(A - BK_k),$$

where the backward recursion is initialized with $P_t = I$.

The practice of combining the two desiderata into one performance index is somewhat arbitrary. There is no fundamental justification for doing so, other than the need to settle on one function to be minimized (it is not generally possible to find one control history that will simultaneously minimize both e_1 and e_2). Forming the weighted sum of the two criteria is also arbitrary. One might just as well consider minimizing the product or, for that matter, any number of functions of e_1 and e_2. The sum is used primarily out of tradition and convenience.

Even after settling on the sum, however, there is still a subtle point that is not usually discussed by control theorists – the issue of inter-utility comparisons. Although individual utilities are invariant with respect to positive affine transformations, utilities cannot be compared with each other unless they are expressed in the same units. Thus, we might assume that R is a units conversion factor between the units of u and the units of $\mathbf{x}$. In practice, however, this is not usually the case. Instead, R is treated as a design parameter and is adjusted to ensure an acceptable compromise between the amount of energy consumed and the size of the error. It is evident that what "optimality" brings to this solution, aside from the aura of respectability, is that it provides a systematic solution technique. At the end of the day there is no pretense that the solution that is ultimately selected is the very best one possible. It seems that, even with the most basic and elegant of optimization problems, one cannot completely escape some vestige of the *ad hoc*.

Let us now generate satisficing solutions to the quadratic linear regulator problem. Our approach is fundamentally different from the optimization approach. Rather than identifying the unique "best" solution, we focus attention on eliminating poor solutions. We do this by defining the appropriate selectability and rejectability functions and applying the PLRT.

There is another significant difference between satisficing and optimization. Under satisficing, we are not looking for a global solution that is the single best option out of all that are available. Instead, we restrict our attention to a limited temporal extent. This approach consists of implementing a feedback controller through a series of repeated open-loop calculations based on the instantaneous state. For a discrete-time receding horizon of length d, the next d values, $\{u_k, \ldots, u_{k+d-1}\}$, are computed as functions of the current state, $\mathbf{x}_k$. The control u_k is implemented, producing a state $\mathbf{x}_{k+1}$, the horizon is shifted forward one time unit, and the process is repeated. For example, a one-step control horizon ($d = 1$) would require the design of only u_k, the control for the current time increment. For $d = 2$, both u_k and u_{k+1} are required. For this problem, let us set $d = 1$ and look only one step ahead by computing the satisficing set of controls only for time k.

We begin by setting upper and lower limits on the control variable. This step is not necessary with the optimal approach, but it is certainly in conformity with reality. Every practical control problem will have limited control authority. Let $U_m = (-u_m, u_m)$ be a finite open interval representing the range of admissible controls.

To calculate selectability, observe that perfect regulation occurs when $\mathbf{x} = 0$. Viewing $k + 1$ as the terminal time, the cost associated with this time is

$$\Phi(u, \mathbf{x}_k) = \mathbf{x}_{k+1}^T \mathbf{x}_{k+1}$$
$$= [A\mathbf{x}_k + Bu]^T [A\mathbf{x}_k + Bu].$$

Now define the function

$$g_S(u, \mathbf{x}_k) = \sup_{v \in U_m} \{\Phi(v, \mathbf{x}_k)\} - \Phi(u, \mathbf{x}_k)\}, \tag{4.2}$$

since we desire control values that make Φ small in order to have high selectability. To determine rejectability, observe that the resource is energy, and, irrespective of the goal, we wish to reject values of u that consume power. Thus, at time k the rejectability must be proportional to the power, given by

$$g_R(u) = Ru^2. \tag{4.3}$$

Because g_S and g_R are quadratic in u for fixed $\mathbf{x}_k$, it is possible to identify the maximum and the minimum of these local functions

$$u_{\mathfrak{E}_S} = \arg\max_{u \in U_m} g_S(u, \mathbf{x}_k),$$
$$u_{\mathfrak{E}_R} = \arg\min_{u \in U_m} g_R(u).$$

It is sufficient to allocate all of the selectability and all of the rejectability to the equilibrium set. It is easily shown that, for quadratic performance indices, the boundaries of the equilibrium set are determined by the maximum value of g_S and the minimum value of g_R. Let

$$u_* = \min\{u_{\mathfrak{E}_S}, u_{\mathfrak{E}_R}\},$$
$$u^* = \max\{u_{\mathfrak{E}_S}, u_{\mathfrak{E}_R}\},$$
$$U = [u_*, u^*] = \mathfrak{E}$$

and assume $u_m \gg |u_*|$ and $u_m \gg |u^*|$ so that the boundaries of U_m do not affect these values.

Since $g_S(u, \mathbf{x}_k)$ is quadratic in u, it achieves its maximum at $u_{\mathfrak{E}_L}$ when restricted to $\mathfrak{E}$. Hence,

$$p_S(u; \mathbf{x}_k) = \frac{g_S(u, \mathbf{x}_k)}{G_S(\mathbf{x}_k)}, \tag{4.4}$$

where the normalizing term is given by

$$G_S(\mathbf{x}_k) = \int_{u_*}^{u^*} g_S(v; \mathbf{x}_k)dv.$$

Similarly, since $g_S(u, \mathbf{x}_k)$ is quadratic, it assumes its unique maximum at $u_{\mathfrak{E}_R}$. Thus,

$$p_R(u) = \frac{g_R(u)}{G_R(\mathbf{x}_k)} \qquad (4.5)$$

where the normalizing term is given by

$$G_R(\mathbf{x}_k) = \int_{u_*}^{u^*} g_R(v)dv.$$

The most selectable and least rejectable controls are given by

$$u_S = u_{\mathfrak{E}_S} = -[B^{\mathrm{T}}B]^{-1}B^{\mathrm{T}}A\mathbf{x}_k, \qquad (4.6)$$

$$u_R = u_{\mathfrak{E}_R} = 0. \qquad (4.7)$$

Because p_S is concave and p_R is convex, $\mathfrak{S}_q$ is convex (see Theorem 4.2), which implies that all possible satisficing equilibrium controls may be obtained via convex combinations of u_R and u_S. For $\lambda \in [0, 1]$ define $u_\lambda = \lambda u_R + (1 - \lambda)u_S$. For $\lambda \approx 0$, the control tends to reduce the accumulated cost at the expense of large terminal error, and for $\lambda \approx 1$, the control tends to reduce the terminal error at the expense of the accumulated cost. Thus, λ is a design parameter for a synthesizing procedure. There exists a λ_D such that the most discriminating control is given by $u_D = u_{\lambda_D}$. The most discriminating control u_D can be calculated directly in a manner similar to the calculations of u_S and u_R (i.e., maximizing $p_S(u, \mathbf{x}_k) - qp_R(u)$ with respect to u), yielding

$$u_k = -[B^{\mathrm{T}}B + q'R]^{-1}B^{\mathrm{T}}A\mathbf{x}_k,$$

where

$$q' = q\frac{G_S(\mathbf{x}_k)}{G_R(\mathbf{x}_k)}. \qquad (4.8)$$

We may define the satisficing gain as

$$\mathcal{K}_k = [B^{\mathrm{T}}B + q'R]^{-1}B^{\mathrm{T}}A,$$

so that $u_k = -\mathcal{K}_k\mathbf{x}_k$. The satisficing control for the linear quadratic regulator is a state-feedback control and has a structure similar to the optimal feedback control. Because the gain $\mathcal{K}_k$ is a function of the state $\mathbf{x}_k$ (via q'), however, the satisficing feedback is not linear.

It may be useful to discuss some of the differences between the optimal and the satisficing solutions.

- The optimal solution is a global solution. It requires knowledge of the system model for the full extent of the problem. If the system model changes unpredictably, the

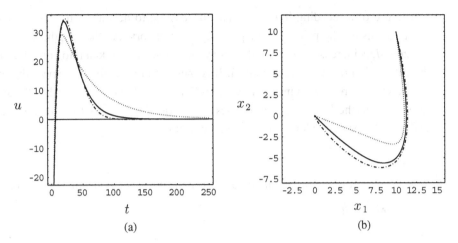

Figure 4.3: Performance results for the single-agent linear regulator problem: (a) control history, (b) phase plane.

solution is invalidated. The satisficing solution is a local solution. It requires knowledge of the system only for the present. If the system model changes unpredictably, the solution will adjust.

- The optimal solution guarantees convergence. The satisficing solution does not.
- The optimal solution requires that the model be linear. The satisficing solution does not exploit linearity. For linear systems, the optimal solution is more computationally efficient than the satisficing solution. For nonlinear systems, however, closed-form optimal solutions are generally not available and the computational burden of approximately optimal solutions may be significant. The computational burden for the satisficing solution is independent of linearity.

To compare the optimal and satisficing control policies, let us use an unstable second-order linear time-invariant system example taken from Kirk (1970). Let

$$A = \begin{bmatrix} 0.9974 & 0.0539 \\ -0.1078 & 1.1591 \end{bmatrix}, \quad B = \begin{bmatrix} 0.0013 \\ 0.0539 \end{bmatrix}, \quad R = 0.05.$$

The resulting control histories are plotted in Figure 4.3(a), with the solid curves representing the optimal control and the other two curves representing the one-step satisficing control for two different values of boldness. The optimal trajectory is shown in Figure 4.3(b) as the solid curve, and the solution of the one-step satisficing controller is displayed by the other two curves using two different boldness values. For the satisficing control, u_D, the most discriminating control, was employed. The optimal cost for this problem is $J_{opt} = 1803$, and the cost for the one-step controllers are $J_1 = 1973$, $J_{1.6} = 1818$. Thus, performance degrades approximately 10% when $q = 1$ and less than 1% when $q = 1.6$ when if a satisficing solution is employed even with a planning horizon of $d = 1$.

The use of a receding horizon makes it possible to develop a tractable satisficing controller using the principles of praxeic utility theory. As the length of the control horizon, d, is increased, more of the future state values are taken into consideration for the calculation of the current control. It is therefore reasonable to expect that performance will improve with increasing d. The following theorem establishes the stronger result that, in the limit as d approaches t_f, the quadratic regulator satisficing control will actually be equivalent to the optimal control.

Theorem 4.4

(Consistency of quadratic regulator) *For the deterministic quadratic regulator problem*

$$\mathbf{x}_{k+1} = A\mathbf{x}_k + Bu_k, \quad k = 0, 1, \ldots, t - 1,$$

$$J(u_0, u_1, \ldots, u_{t-1}) = \mathbf{x}_t^T \mathbf{x}_t + \sum_{k=0}^{t-1} Ru_k^2.$$

If the control horizon spans the full extent of the problem, that is, for $d = t$, then the most discriminating satisficing control is identical to the optimal control.

PROOF

The most discriminating control, denoted $\mathbf{u} = \{u_0, \ldots, u_{t-1}\}$, is

$$\mathbf{u} = \arg \sup_{\mathbf{v} \in \mathbb{R}^t} \{p_S(\mathbf{v}) - qp_R(\mathbf{v})\},$$

where $\mathbb{R}^t = \mathbb{R} \times \cdots \times \mathbb{R}$, t times. From (4.4), (4.5), and (4.8),

$$\arg \sup_{\mathbf{v} \in \mathbb{R}^t} \{p_S(\mathbf{v}) - qp_R(\mathbf{v})\} = \arg \sup_{\mathbf{v} \in \mathbb{R}^t} \{g_S(\mathbf{v}) - q'g_R(\mathbf{v})\}.$$

Also, from (4.2) and (4.3),

$$\arg \sup_{\mathbf{v} \in \mathbb{R}^t} \{g_A(\mathbf{v}) - q'g_L(\mathbf{v})\} = \arg \inf_{\mathbf{v} \in \mathbb{R}^T} \left\{ \mathbf{x}_t^T \mathbf{x}_t + q' \sum_{t=0}^{t-1} Ru_k^2 \right\}.$$

Thus,

$$\mathbf{u} = \arg \inf_{\mathbf{v} \in \mathbb{R}^t} \left\{ \mathbf{x}^T(t)\mathbf{x}(t) + q' \sum_{k=0}^{t} +Rv_k^2 \right\}.$$

For $q' = 1$, this is exactly the optimal quadratic regulator solution. ☐

Theorem 4.4 also holds for nonlinear systems (see Goodrich et al. (1998)).

5 Uncertainty

Solum certum nihil esse certi.
The only certainty is uncertainty.

Pliny the Elder
Historia Naturalis, Bk ii, 7 (1535)

The will is infinite and the execution confined . . .
the desire is boundless and the act a slave to limit.

William Shakespeare
Troilus and Cressida, Act 3 scene 2 (1603)

That decision makers can rarely be certain is obvious. The question is, what are they uncertain about? The most well-studied notion of uncertainty is *epistemic uncertainty*, or uncertainty that arises due to insufficient knowledge. Such uncertainty is usually attributed to randomness or imprecision. The effect of epistemic uncertainty is to increase the likelihood of making erroneous decisions. The worst thing one can do in the presence of epistemic uncertainty is to ignore it. It is far better to devise models that account for as much of the uncertainty as can be described. Consequently, a large body of theories of decision making in the presence of less than complete information has been developed over many decades. I do not propose to give an exhaustive treatment of the way epistemic uncertainty is accounted for under classical decision-making paradigms, but instead summarize two ways to deal with uncertainty. The first is the classical Bayesian approach of calculating expected utility, and the second is the use of convex sets of utility functions.

There is a second type of uncertainty, which I term *praxeic uncertainty*. Praxeic uncertainty does not deal with the insufficiency of knowledge. The option that is chosen may be made with complete knowledge – correct in every respect. Evaluation of the correctness of a solution, however, is not the whole story. We may also be interested in how difficult it is for the decision maker to arrive at its choice or set of choices. Praxeic uncertainty is an ecological issue, rather than a nescience issue, and deals with the innate ability of the decision maker to accomplish its task. The effect of praxeic uncertainty is to reduce the functionality and sensitivity of the decision maker. There are two manifestations of praxeic uncertainty. The first, which I call equivocation, deals with the fitness, or suitability, of the decision maker to operate effectively in the environment in which it must function. The second manifestation of praxeic uncertainty deals with what I term quasi-invariance, which deals with a fundamental arbitrariness of the decision problem that results, not from the lack of information, but from the arbitrariness of the conventions that are used to ground the decision-making criteria. This type of uncertainty deals more with questions of sensitivity than with correctness.

5.1   Bayesian uncertainty

With many decision-making models there exists a set of parameters, called the state of Nature, denoted Θ, that influences the preferences of the decision maker and generates a family of utility functions parameterized by $\theta \in \Theta$. The Bayesian approach to this situation is to view θ as a random variable and ascribe a probability distribution to it, called the *prior*, denoted P_C. The prior may be used to calculate the expected value of the utility function.

In the contexts of epistemic utility theory and praxeic utility theory, both the selectability and rejectability measures may be functions of the state of Nature. The expected epistemic utility function given by (3.6) becomes

$$\pi(A|\vartheta) = \sum_{u \in A} [p_S(u|\vartheta) - q p_R(u|\vartheta)],$$

where $p_S(\cdot|\vartheta)$ and $p_R(\cdot|\vartheta)$ are now indexed by[1] $\vartheta \in \Theta$. Note that in Chapter 3 we referred to π as expected epistemic utility, where the expectation was taken with respect to error avoidance, p_S. Now, however, we have added an additional level of complexity by considering the state of Nature, which means that the original expected epistemic utility is now a random variable, since it is a function of the random variable θ. We may average out the randomness by taking the expectation of the original expected epistemic utility with respect to the prior, yielding

$$E_\theta \pi(A) = \sum_{u \in A} [\bar{p}_S(u) - q \bar{p}_R(u)],$$

where E_θ is the mathematical expectation operator with respect to the random variable θ, and

$$\bar{p}_S(u) = \sum_{\vartheta \in \Theta} p_S(u|\vartheta) p_C(\vartheta),$$

$$\bar{p}_R(u) = \sum_{\vartheta \in \Theta} p_R(u|\vartheta) p_C(\vartheta),$$

---

[1] To distinguish between the random variable and the values that it assumes, let us use θ to denote the former and ϑ to denote latter. Thus, we may make sense of such statements as $P_C[\theta = \vartheta]$.

where p_C is the credal probability that characterizes the state of Nature. Under the Bayesian approach, $\bar{p}_S$ and $\bar{p}_R$ would be used to form the PLRT.

Example 5.1 Let us revisit Example 3.2, where Melba is trying to decide which message was sent. If Milo's favorite team is in town, there is a high likelihood that he would want to invite her to it, and that information would cause her to reconsider her choice of p_S. The trouble is, she does not know whether or not that team is in town. The state of Nature for this problem is, therefore, $\Theta = \{T, A\}$, where T stands for the team being in town and A stands for the team being away.

Suppose, if the team were in town, that Melba's belief state regarding the messages would be described by the probability assignment

$p_S(\text{M-1}|T) = 0.25,$

$p_S(\text{M-2}|T) = 0.50,$

$p_S(\text{M-3}|T) = 0.25,$

and, if the team were away, her probability assignment would be

$p_S(\text{M-1}|A) = 0.5,$

$p_S(\text{M-2}|A) = 0.0,$

$p_S(\text{M-3}|A) = 0.5.$

Melba's prior is of the form $p_C(T) = \alpha$, $p_C(A) = 1 - \alpha$, where $0 \le \alpha \le 1$, and the expected selectability mass function is

$\bar{p}_S(\text{M-1}) = 0.5 - 0.25\alpha,$

$\bar{p}_S(\text{M-2}) = 0.5\alpha,$

$\bar{p}_S(\text{M-3}) = 0.5 - 0.25\alpha.$

Using the values for rejectability as given in (3.9) and setting $q = 1$, Melba would never reject (M-3) and would retain both (M-2) and (M-3) if $\alpha \ge 0.6$.

This approach is in the tradition of what is often termed "decision making under risk." According to this approach, the state of epistemic uncertainty is completely characterized by a numerically definite prior distribution. This assumption fits well with the von Neumann–Morgenstern concept of maximizing expectations, and the decision maker simply maximizes its expected utility by averaging over all forms of randomness, including the state of Nature. Another school of thought regarding epistemic uncertainty is to deny the assertion that the state of Nature can be appropriately modeled as a random variable and to assert that epistemic uncertainty should not be characterized by probability distributions. Adherents of this view often propose that, in the presence of such uncertainty, the decision maker should adopt an extremely conservative approach by viewing Nature as an adversary. Under this assumption, decision making can be viewed as a zero-sum game between the decision maker and

Nature, and the decision maker should choose the option that minimizes the maximum damage that Nature can inflict.

5.2   Imprecision

A decision maker who does not have a clear understanding of how to specify preferences regarding the conservation of resources is in a state of value conflict. A decision maker who does not have a clear understanding of how to specify preferences regarding the goals to be pursued is in a state of operational conflict. Strict Bayesian doctrine requires that the decision maker, even if by fiat, declare all conflicts to be resolved. The device used to effect this resolution is to impose a numerically precise prior. But rational decision makers cannot be expected to be so decisive in their assessment of goals and values that they are always willing to provide precise numerical specifications for their utilities. Circumstances do not always provide a basis for resolving conflicts before making a decision. Wisdom dictates that, when the decision maker is not in possession of numerically definite utilities, the decision maker should be willing to suspend judgment between all rational decisions that are seriously possible given the data that are in its possession.

X may be conflicted between multiple informational valuations and may not be willing to decide between, say, p_R and p'_R, both of which capture aspects of informational value but are not consistent. One way to address the existence of such conflicts is to form a family of rejectability functions composed of all convex combinations of p_R and p'_R; that is, the family of functions of the form $\{p_R^\beta = \beta p_R + (1 - \beta)p'_R : \beta \in [0, 1]\}$ as representing the range of informational values of the decision maker. The parameter β represents X's compromise between the conflicting informational valuation systems.

If the set of rejectability functions used to represent X's informational values is a convex set, a condition of informational convexity occurs. Let $\mathcal{G}$ denote a set of rejectability functions. We say that $\mathcal{G}$ is X's informational state, and assume that $\mathcal{G} \neq \emptyset$, thus ensuring that X is never completely indifferent to the outcome of the decision problem. Let us assume that $\mathcal{G}$ is convex; that is, if $p_R \in \mathcal{G}$ and $p'_R \in \mathcal{G}$ then $\beta p_R + (1 - \beta)p'_R \in \mathcal{G}$ for all $\beta \in [0, 1]$. We shall say that a condition of informational uniqueness obtains if $\mathcal{G}$ consists of a singleton set, the rejectability function $\{p_R\}$.

I do not attempt to prove that conservational convexity is the best way to deal with unresolved conflict. But, by admitting any convex combination of rival rejectability functions to be a rejectability function also, the decision maker does not rule out any potential resolutions of the conflict. By permitting conflicting informational values to co-exist, there remains the possibility that future data will permit an ultimate resolution.

Suppose, rather than Melba being completely numerically definite regarding her rejectability function and accepting the rejectability function provided in (3.9), she also is considering the alternative rejectability function

$$p'_R(\text{M-1}) = 0.30,$$
$$p'_R(\text{M-2}) = 0.65, \tag{5.1}$$
$$p'_R(\text{M-3}) = 0.05.$$

Melba can express this imprecision by the linear combination

$$p_R^\beta(\text{M-1}) = \beta p_R(\text{M-1}) + (1-\beta)p'_R(\text{M-1}) = 0.30 + 0.30\beta,$$

$$p_R^\beta(\text{M-2}) = \beta p_R(\text{M-2}) + (1-\beta)p'_R(\text{M-2}) = 0.65 - 0.30\beta,$$

$$p_R^\beta(\text{M-3}) = \beta p_R(\text{M-3}) + (1-\beta)p'_R(\text{M-3}) = 0.05,$$

where $0 \le \beta \le 1$. The rejectability function p_R^β reflects the situation that Melba is not completely sure about how to apportion her unit of informational value between (M-1) and (M-2), so she parameterizes this specification with β, but she is quite sure that her informational value for rejecting the marriage question is 0.05, so she will trade off informational value between only (M-1) and (M-2). This formulation provides her with a capability to hedge her feelings and to analyze how the epistemic uncertainty in her informational value assessments propagates into her decision-making capabilities.

Melba may apply the expected epistemic utility rule in this situation. If she uses the same uniform selectability as before and takes $q = 1$, she finds that she now has a *set* of options to consider:

reject (M-2) only   for $\beta \ge 0.1111$,
reject (M-1) and (M-2)   for $\beta < 0.1111$.

A decision maker may not only be conflicted regarding informational values, but it also may be conflicted regarding its assessments of how to achieve its goals. The selectability state of X is the set of all selectabilities that are regarded by X as seriously possible descriptions of the inquiry under investigation. Let $\mathcal{B}$ denote the selectability state for X relative to a given inquiry. Suppose $p_S \in \mathcal{B}$ and $p'_S \in \mathcal{B}$ are two possible selectabilities; X is unable to choose between them, and thereby displays its ignorance. As with informational value, one way to address this problem is to relax the requirement for a numerically definite selectability and consider all convex combinations of p_S and p'_S.

If the selectability state is closed under all finite convex combinations, then a condition of selectability convexity occurs. That is, if $p_S \in \mathcal{B}$ and $p'_S \in \mathcal{B}$, then $p_S^\alpha = \alpha p_S + (1-\alpha)p'_S \in \mathcal{B}$ for $\alpha \in [0, 1]$. The decision maker's ignorance may be expressed through a convex selectability state. If X has sufficient knowledge to warrant the specification of a single selectability, then a condition of selectability uniqueness exists, and the selectability state contains exactly one selectability. Convexity permits the decision maker to incorporate its ignorance into the decision problem by relaxing the requirement that the distribution be numerically definite. Instead, it specifies a convex region in the space of logically possible selectability functions to be considered as serious possibilities.

Let us now establish an important convexity property. Suppose the set of prior measures forms a convex set. The following theorem (Levi, 1980) guarantees that the set of posterior measures is also convex.

Theorem 5.1

Let B be a convex set of unconditional unit-normed measures and define the set of conditional measures of the form

$$B_c = \left\{ \xi_G(\cdot) = \frac{\xi(\cdot \cap G)}{\xi(G)} : G \in \mathcal{F}, \xi(G) > 0 \,\forall \xi \in B \right\}.$$

Then B_c is convex.

PROOF

Let $\xi \in B$, $\xi' \in B$, and $\alpha \in [0, 1]$. For any $G \in \mathcal{F}$ such that $\xi(G) > 0$ and $\xi'(G) > 0$, and for any set $E \in \mathcal{F}$, we form the conditional measures

$$\xi_G(E) = \frac{\xi(E \cap G)}{\xi(G)}, \quad \xi'_G(E) = \frac{\xi'(E \cap G)}{\xi'(G)},$$

both of which lie in B_c. For the convex combination

$$\xi^\alpha(E) = \alpha\xi(E) + (1 - \alpha)\xi'(E),$$

the conditional measure

$$\xi^\alpha_G(E) = \frac{\xi^\alpha(E \cap G)}{\xi^\alpha(G)}$$

also lies in B_c. The theorem will be proved if there exists $\beta \in [0, 1]$ such that

$$\xi^\alpha_G(E) = \beta\xi_G(E) + (1 - \beta)\xi'_G(E) \tag{5.2}$$

holds. Direct substitution shows that (5.2) holds for

$$\beta = \frac{\alpha\xi(G)}{\xi^\alpha(G)}.$$

Since $0 \le \beta \le 1$, the theorem is proved.   □

Meanwhile, Melba is still having problems determining which question was asked, since she realizes that she is informationally conflicted. As she ponders, she realizes that she is also conflicted with respect to her selectability state, since she arbitrarily specified the uniform distribution for her selectability state and she had no data or other evidence to justify doing so. Being conservative, she reasons that her prior probability is more appropriately modeled as a convex set of the form

$$p_S(\text{M-1}) = \kappa,$$

$$p_S(\text{M-2}) = \lambda, \tag{5.3}$$

$$p_S(\text{M-3}) = 1 - \kappa - \lambda,$$

where κ and λ are such that

$$\kappa_{\min} \leq \kappa \leq \kappa_{\max},$$

$$\lambda_{\min} \leq \lambda \leq \lambda_{\max},$$

$$\kappa + \lambda \leq 0.9,$$

where $\kappa_{\min}, \kappa_{\max}, \lambda_{\min}$ and $\lambda_{\max}$ are all in the open interval $(0, 1)$ with $\kappa_{\min} < \kappa_{\max}$ and $\lambda_{\min} < \lambda_{\max}$. Equations (5.3) represents Melba's selectability state. Figure 5.1 illustrates this selectability simplex for $\kappa_{\min} = \lambda_{\min} = 0.05$ and $\kappa_{\max} = \lambda_{\max} = 0.9$.

As Melba contemplates how to resolve her problem, she recalls that Milo recently withdrew a sum of money from his bank. She reasons that, were he to propose marriage, he might use the money to purchase an engagement ring, but he also might need it to purchase ball game tickets, which are moderately expensive. Letting W denote the act of withdrawing the money, she determines conditional probabilities for withdrawing the money, given the three possibilities, are $P(W|\text{M-3}) = 0.74$, $P(W|\text{M-2}) = 0.25$, and $P(W|\text{M-1}) = 0.01$. These probabilities, though numerically precise, reflect local, rather than global, behavior and isolate consideration to the decision problem at hand. The localized nature of this context makes it easier to specify numerically precise

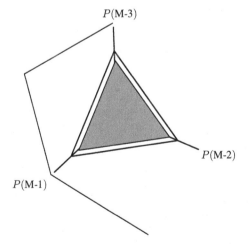

Figure 5.1: The prior selectability simplex for Melba.

probabilities than is the case for the general problem of specifying the selectability mass function.

Applying Bayes theorem, Melba calculates that her posterior probability function is

$$P(\text{M-1}|W) = \frac{0.01\kappa}{0.74 - 0.73\kappa - 0.49\lambda},$$

$$P(\text{M-2}|W) = \frac{0.25\lambda}{0.74 - 0.73\kappa - 0.49\lambda},$$

$$P(\text{M-3}|W) = \frac{0.74(1 - \kappa - \lambda)}{0.74 - 0.73\kappa - 0.49\lambda}.$$

Melba's posterior selectability state is

$$\frac{0.01\kappa_{\min}}{0.74 - 0.73\kappa_{\min} - 0.49\lambda_{\min}} \leq \kappa \leq \frac{0.01\kappa_{\max}}{0.74 - 0.73\kappa_{\max} - 0.49\lambda_{\min}},$$

$$\frac{0.25\lambda_{\min}}{0.74 - 0.73\kappa_{\min} - 0.49\lambda_{\min}} \leq \lambda \leq \frac{0.25\lambda_{\max}}{0.74 - 0.73\kappa_{\min} - 0.49\lambda_{\max}},$$

$$\kappa + \lambda \leq 0.9.$$

Figure 5.2 illustrates this selectability state for $\kappa_{\min} = \lambda_{\min} = 0.05$ and $\kappa_{\max} = \lambda_{\max} = 0.9$.

Using the same maximum and minimum values for κ and λ as before, the Melba's posterior selectability state becomes

$$0.0007 < \kappa < 0.1539,$$

$$0.0184 < \lambda < 0.8571.$$

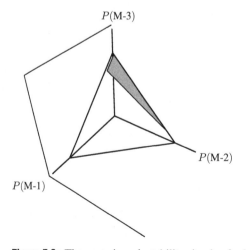

Figure 5.2: The posterior selectability simplex for Melba.

5.3   Equivocation

Human decision makers often make qualitative assessments of the difficulty, in terms of stress or tension, encountered in making decisions. Even if such knowledge does not have a direct bearing on their immediate decisions, an appreciation of the ecological difficulty involved in forming the decision is an important aspect of the decision-making experience. A decision maker need not possess anthropomorphic qualities, however, to assess the difficulty of making decisions. Nor do I propose to endow an artificial decision maker with some sort of ersatz anthropomorphic capability. With the comparative paradigm provided by praxeic utility theory, however, it is possible to evaluate attributes of the decision problem that correspond more to the functionality and fitness of the decision maker than to its ability to choose correctly.

5.3.1   Attitude

Are decisions easily made and implemented, or do they tax the capabilities of the decision maker in fundamental ways? Such assessments are not a typical undertaking of classical decision theory based on substantive rationality. Maximizing expectations has no need to concern itself with issues such as "difficulty" (except perhaps as decisions become "too difficult" in the sense that they exceed the decision maker's computational ability). Nevertheless, choices are not all of equal difficulty.

Consider Wendle's situation in Example 3.4. Suppose the probability of rain is $\beta = 0.7$. Wendle's only rational choice under the maximum expectations paradigm is to go to the game, even though it is highly likely that he will get wet. He makes this decision because he strongly prefers the ball game to the museum. In this case, Wendle is a gambling man and is prepared to live with the consequences, rain or shine.

Obviously, going to the game when the sun is likely to shine is an emotionally easy decision, whereas going to the game when the probability of rain is moderately high is emotionally more difficult. If the probability of rain is high, then Wendle's emotional state might cause him to question, up to the moment of truth, his desire to go to the game. But any real doubt would, according to substantive rationality, simply cause him to adjust his preferences, and hence his utilities, and to reassess his decision. We must assume that issues such as the future state of Wendle's health, his travel arrangements, the time commitment involved, and so forth, were all carefully considered when Wendle established his preferences and generated his utilities and therefore are not subject to change. We also assume that unforeseen situations, such as accidents, that might affect Wendle's will or ability to carry out his rationally conceived decision, have been accommodated in the evaluation of his expected utility.

As an expected utility maximizer, Wendle's decision mechanism does not brook any vacillation on the part of his will or ability. There is no room for equivocation. There is no way for Wendle to second-guess himself without sacrificing his rationality. In fact, there is no way for him even to evaluate the difficulty of his decision, should he wish to do so. His utility function cannot provide him with any such information – it simply and dispassionately rank-orders his preferences and his decision mechanism instructs him to act according to that ranking. Even if he concocts some measure of difficulty, doing so is simply a *post factum* exercise that does not change his decision-making procedure, provided that he remains committed to utility maximization.

Ace is also a gambling man who is prepared to live with the consequences of his actions. Acting according to praxeic utility theory, however, affords him some means of assessing the difficulty of making his decision. Unlike Wendle, who has only one utility function to consider, Ace has two utility functions that are expressly designed to make comparisons. Not only may he compare them, but he may manipulate them in various ways with the hope of identifying selectability/rejectability relationships that will characterize his "attitude" or "disposition" toward the problem at hand.

The difficulty for Ace (and for Wendle, too) does not derive from an epistemic source – it is not due to epistemic uncertainty about the weather. Indeed, there is only moderate uncertainty in that regard – with $\beta = 0.7$, it is more likely to rain. If the probability of rain were to increase to, say, $\beta = 0.74$, the epistemic uncertainty regarding the weather would be less, but the decision would be even harder. *The reason the decision is difficult is that there is a fundamental conflict between the two desiderata that generate his two utility functions.* Doing things in the interest of avoiding failure (staying dry), as characterized by p_S, is in opposition to the conservation of his resources (personal enjoyment), as characterized by p_R.

By employing two utilities, rather than only one, it is possible to analyze them in order to establish the compatibility of the two sets of preferences that exist. If these preferences are compatible, in that options that conserve resources also avoid failure, then the decision maker is in a fortunate situation. If the preferences are incompatible, in that options that avoid failure also are highly consuming of resources, then the decision maker is fundamentally conflicted. These situations illustrate the attitudes, or dispositions, of the decision maker.

The literature of substantive rationality is devoid of discussions of the attitudes or dispositions of the decision maker who is assumed to be dispassionate. It is simply doing what should be done according to rationality alone. Any attitudes or feelings, should they even exist (and they need not), are completely irrelevant. Furthermore, to attribute any anthropomorphic characteristics to such a decision maker might be seen as nothing more than a concocted story line that is of marginal value, if not completely misleading.

Attitudes, however, cannot be dismissed so easily, even in the sterile environment of expectations maximization. Like it or not, restlessness, insatiability, and intemperance are at least tacitly imposed even on an intransigent optimizer. Such attitudes can lead to inflexible behavior. Especially in multi-agent environments, which I will consider in

Chapter 6, they may lead to cynical and even antisocial behavior (to import additional anthropomorphic, but highly relevant, terms).

Optimality does not ensure that the decision is satisfactory to the decision maker.[2] If none of the available options achieve the goal adequately, or if all available options result in extremely unpleasant side effects, adopting the optimal solution may simply be making the best of a bad situation (whatever comfort that knowledge affords).

It is fortunate if an option that conserves resources (low rejectability) also avoids failure (high selectability). In this circumstance, a decision maker is content. Many interesting decision problems, however, are such that options that might be chosen in the interest of avoiding failure are expensive, hazardous, or have other undesirable side effects. A decision maker in this situation is conflicted. Contentment and conflict are basic dispositional states that serve as guides to the decision maker's functionality. A situation requiring frequent high-conflict decisions will be a difficult one for the decision maker. Making high-conflict decisions, however, is not a measure of how well the decision maker is performing. It may, in fact, be making good, but unavoidably costly, decisions. It is also true that a high-conflict environment may result in poor performance because the decision maker is simply not resourceful enough to deal adequately with its environment. Such a situation might serve as a trigger to prompt changes in the decision maker, such as activating additional sensors, or seeking more information about the environment through other means, or obtaining therapy in order to improve its behavior. The situation may also trigger a learning mechanism that allows the decision maker to adapt itself better to the environment.

We may gain insight into the contentedness/conflict disposition of a decision maker by examining its selectability and rejectability functions. For this analysis, we concentrate on the case involving only finitely many options (the theory also extends to the case of a continuum of options, but little additional insight is gained by so doing). Since selectability and rejectability are mass functions, it may be useful to appropriate some additional mathematical machinery of probability theory to aid in interpreting them. To this end, we introduce the concept of entropy.

Definition 5.1
The **entropy** of a mass function p is

$$H(p) = - \sum_{u \in U} p(u) \log_2 p(u). \qquad \Box$$

It is easily shown (for example, see Ash (1965) or Cover and Thomas (1991)) that entropy is non-negative; that is, $H(p) \geq 0$ for all mass functions p defined over U. Also, entropy is concave; that is, if p_1 and p_2 are two mass functions and $0 \leq \lambda \leq 1$,

---

[2] As a colleague of mine once noted, "When you travel on company business, the very best accommodations are barely adequate."

then $H[\lambda p_1 + (1 - \lambda)p_2] \geq \lambda H(p_1) + (1 - \lambda)H(p_2)$. It is customary to use base 2 logarithms; doing so results in entropy units of *bits*, a coined word for binary digits. Although such units are not relevant to our analysis, we will continue with the custom of using base 2 logarithms.

To understand entropy, let us briefly digress and discuss its use in its conventional setting, that of Shannon information theory. In this context, the epistemic uncertainty of an outcome is the degree to which it is unlikely to occur. For a given probability mass function p we define what is sometimes called the **surprise function** (see, e.g., Ross (2002)),

$$S_p(u) = \log_2 \frac{1}{p(u)} = -\log_2 p(u).$$

The surprise function is a measure of the epistemic uncertainty associated with u and quantifies the degree to which the outcome u can be anticipated. For example, suppose $p(u) \approx 1$, so $S_p(u) \approx 0$. Intuitively, since the occurrence of u is highly probable, we can predict with considerable confidence that the outcome of an experiment governed by p will be u, and if the experiment were to be repeated many times, we would rarely be surprised by the occurrence of any other outcome. In other words, the epistemic uncertainty in the outcome will be reduced only slightly by actually performing the experiment. Now, suppose $p(u') \approx 0$, so u' is highly unlikely to occur. Then $S_p(u')$ is large, indicating that great epistemic uncertainty is associated with predicting the occurrence of that event, or, equivalently, epistemic uncertainty in the outcome is greatly reduced by the occurrence of u'.

Entropy is the average value of epistemic uncertainty over all $u \in U$; that is, it is the average amount of surprise one receives upon learning the outcome of an experiment. There are two equivalent interpretations of this quantity. On the one hand, $H(p)$ can be viewed as a measure of the average epistemic uncertainty in the outcome of an experiment governed by p *before* it is conducted. On the other hand, it is a measure of the average reduction in epistemic uncertainty *after* the experiment has been conducted. Putting this latter interpretation slightly differently, entropy is the average increase in epistemic certainty as a result of conducting the experiment.

To appreciate entropy in the context of praxeic utility theory, we require interpretations of entropy for both selectability and rejectability in ways that are analogous to the usual Shannon interpretation. This will involve two manifestations of praxeic uncertainty, which we term praxeic uncertainty of the first kind and praxeic uncertainty of the second kind.

Let us first consider praxeic uncertainty of the first kind, and view the expediency of an option as the degree to which it leads to the avoidance of failure. Then **inexpediency**, the degree to which an option leads to failure, is the notion of praxeic uncertainty associated with selectability. It is analogous to epistemic uncertainty, the degree to which an outcome is unlikely to occur. In this context, the surprise function can be interpreted as a **futility function** which quantifies the degree to which an option fails. To illustrate, if p_S is a selectability mass function and $p_S(u) \approx 1$, then $S_{p_S}(u) \approx 0$, which indicates that, since implementing u is highly selectable, there is little inexpediency associated with doing so. Conversely, suppose $p_S(u') \approx 0$. Then $S_{p_S}(u')$ is large, indicating that great inexpediency is associated with implementing that option. $H(p_S)$ then becomes a measure of the average inexpediency (that is, the average futility) associated with the decision problem before taking action. Equivalently, it is a measure of the average reduction in inexpediency after taking action, or to put it slightly differently, it is the average increase in the degree to which success (praxeic certainty of the first kind) is achieved as a result of taking action.

To interpret the entropy of rejectability, let us view expense as the degree to which resources are consumed, and consider its complement – **inexpense** – as praxeic uncertainty of the second kind. In this context, the surprise function can be interpreted as a **frugality function**. If $p_R(u) \approx 1$, then $S_{p_R}(u) \approx 0$, which indicates that u is highly rejectable, thus little inexpense (or, alternatively, very much expense) occurs if a highly rejectable option is implemented. On the other hand, if $p_R(u') \approx 0$ and u' is implemented, then $S_{p_R}(u')$ is large, indicating that great inexpense (little, if any, expense) obtains if u is implemented. $H(p_R)$ is a measure of the average inexpense, or frugality, associated with the decision problem before taking action. Equivalently, it is a measure of the average reduction in inexpense after taking action, or to put it more intuitively, it is the average increase in the consumption of resources (praxeic certainty of the second kind) as the result of taking action.

Let n be the cardinality of the option space U (assumed to be finite). Entropy is maximized by the uniform distribution; that is, if $p^*(u) = \frac{1}{n}$ for all $u \in U$, then $H(p^*) \geq H(p)$ for all mass functions p over U, and has entropy $H(p^*) = \log_2 n$. A near-uniform p_S would generate high average expediency, in that all options would work equally well – but none would perform exceptionally well. A low-entropy p_S would indicate that most of the selectability mass is concentrated on a few options that are conducive to the avoidance of failure. A near-uniform p_R would generate high average inexpense, meaning that all of the options cost the same, and none are inexpensive. A low-entropy p_R indicates that the rejectability mass is concentrated on a few options that consume a disproportionate amount of resource and, consequently, there exists a subset of options that are inexpensive; implementing them will conserve resources.

Definition 5.2

If $p_S(u) = \frac{1}{n}$ (that is, selectability under p_S is equal to the uniform distribution), then the option is **failure neutral**. If the selectability mass function is uniform, then the decision maker's attitude will be one of neutrality in avoiding failure.   □

Definition 5.3

If $p_R(u) = \frac{1}{n}$ (that is, rejectability under p_R is equal to the uniform distribution), then the option is **conservation neutral**. If the rejectability mass function is uniform, then the decision maker's attitude will be one of conservation neutrality.   □

Definition 5.4

If $p_S(u) > \frac{1}{n}$ (that is, the selectability of u under p_S is greater than the selectability of u under the uniform distribution), then u is attractive with respect to performance – u is **expedient**.   □

Definition 5.5

If $p_R(u) > \frac{1}{n}$ (that is, the rejectability of u under p_R is greater than the rejectability of u under the uniform distribution), then u is unattractive with respect to cost or other penalty – u is **expensive**.   □

The relationship between selectability and rejectability permits the definition of four dispositional modes of the decision maker with respect to each of its options. Let U be the set of all possible options.

Definition 5.6

If $u \in U$ is expedient and expensive, then the decision maker will be in a position of desiring to reject, on the basis of cost, an option that is attractive in terms of avoiding failure – it will be in a dispositional mode of **ambivalence** with respect to u. Let U_A be the set of all ambivalent options.   □

Definition 5.7

If $u \in U$ is inexpedient and inexpensive then the decision maker will desire the option on the basis of cost, but will be reluctant to do so because of poor performance in avoiding failure. The decision maker will be in a dispositional mode of **dubiety** with respect to u. Let U_D be the set of all dubious options.   □

Definition 5.8

If $u \in U$ is expedient and inexpensive, then the decision maker is in the position of desiring to implement it – a dispositional mode of **gratification** with respect to u. Let U_G be the set of all gratifying options.   □

Definition 5.9

If $u \in U$ is inexpedient and expensive, then the decision maker will desire to reject it. If the decision maker were compelled to adopt such an option, it would be in a state of frustration. Since, however, not all of the decision maker's options will be in this category, it will always be able to refuse to adopt any such options. We thus say that,

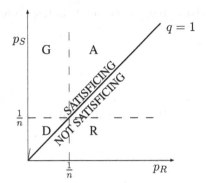

Figure 5.3: Dispositional regions: G = gratification, A = ambivalence, D = dubiety, R = relief.

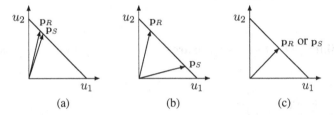

Figure 5.4: Example attitude states for a two-dimensional decision problem.

if u is inexpedient and expensive, then the decision maker is in a dispositional state of **relief** with respect to u. Let U_R be the set of all such options.   □

Clearly,

$$U = U_A \cup U_D \cup U_G \cup U_R.$$

These four modes provide a qualitative measure of the way the decision maker is matched to its task. Gratification and relief are **modes of contentment**, while dubiety and ambivalence are **modes of conflict**. To consider these modes, let us return again to Example 3.4, and note that Ace's attitude with respect to H is always one of relief. His attitude with respect to M is one of ambivalence if $\beta > 0.35$, otherwise it is one of relief. For G, his attitude is one of gratification if $\beta \leq \frac{2}{3}$, one of dubiety if $\frac{2}{3} < \beta \leq \frac{3}{4}$, and one of relief for $\beta > \frac{3}{4}$.

Figure 5.3 illustrates the dispositional regions for $q = 1$ (recall that q is the index of boldness – see Section 3.3). As a further illustration of these modes, Figure 5.4 displays various cases for $n = 2$, a two-dimensional decision problem. In these plots, the diagonal line represents the simplex and the p_S and p_R values are plotted as vectors that lie on the simplex (i.e., the sum of the two components is one). For Figure 5.4(a) the decision maker is dubious with respect to u_1 and ambivalent with respect to u_2, for Figure 5.4(b) the decision maker is gratified with respect to u_1 and relieved with respect to u_2, and for Figure 5.4(c) the decision maker is neutral with respect to either u_1 or u_2.

5.3.2   Figures of merit

Formal expressions to capture some of the features of the qualitative analysis described in Section 5.3.1 utilize two measures that are similar but not identical. I term these two figures of merit *diversity* and *tension*.

Diversity

One important feature of the selectability and rejectability functions is their dissimilarity. To obtain a measure of dissimilarity, we again appeal to the notion of entropy and apply the Kulback–Leibler distance measure.

Definition 5.10
The **Kulback–Leibler (KL) distance measure** of two mass functions, say p_1 and p_2, is given by

$$D(p_1 \| p_2) = \sum_{u \in U} p_1(u) \log_2 \frac{p_1(u)}{p_2(u)}.$$

    □

The KL distance measure is an indication of the relative entropy of two mass functions. $D(\cdot \| \cdot)$ is not a true metric; it is not symmetric and does not obey the triangle inequality. It is, however, non-negative, and it is easily seen that $D(p_1 \| p_2) = 0$ if and only if $p_1(u) = p_2(u)$ for all $u \in U$.

We may apply the KL distance measure to the problem of ascertaining the dissimilarity of the selectability and rejectability functions.

Definition 5.11
The **diversity functional** is:

$$D(p_S \| p_R) = \sum_{u \in U} p_S(u) \log_2 \frac{p_S(u)}{p_R(u)},$$

or, equivalently,

$$D(p_S \| p_R) = -\sum_{u \in U} p_S(u) \log_2 p_R(u) - H(p_S).$$

    □

Small values of the diversity functional are obtained when the selectability and rejectability functions are similar, indicating a condition of potential conflict. If they are identical, then the decision maker is in a position of wishing to reject, on the basis of cost, precisely the options that are in its best interest in terms of performance – an unfortunate condition that may be totally paralyzing.

When $n = 2$ we may express selectability and rejectability mass functions by the vectors $\mathbf{p}_S = [s, 1 - s]$ and $\mathbf{p}_R = [r, 1 - r]$, respectively, as s and r each range over

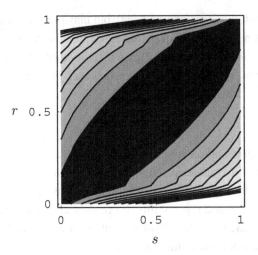

Figure 5.5: The contour plot of the diversity functional for a two-dimensional decision problem.

the unit interval, and the diversity function becomes

$$D(p_S \| p_R) = s \log_2 \frac{s}{r} + (1 - s) \log_2 \frac{1 - s}{1 - r}.$$

Figure 5.5 illustrates a contour plot of the diversity functional for this case. Lighter shading indicates greater diversity. Observe that the diversity value is low for $s \approx r$, but increases as the difference between these values grows.

Returning to Example 3.2, Melba is still in her quandary regarding the message received from Milo. With

$$\mathbf{p}_S = [0.333, 0.333, 0.333]$$

and

$$\mathbf{p}_R = [0.65, 0.35, 0.05],$$

Melba's diversity is $D(p_S \| p_R) = 0.57$. Now suppose that, upon reflection, Melba decides that the institution of marriage is not overwhelmingly more important than the institution of athletics (this decision does not, however, necessarily reflect her personal interest in either of these pursuits). This might cause her to recompute her rejectability to become, say,

$$\mathbf{p}'_R = [0.65, 0.25, 0.15].$$

Her diversity would now be $D(p_S \| p'_R) = 0.20$. Under the former model of rejectability, Melba's satisficing set under unit boldness ($q = 1$) is $\Sigma_q = \{\text{M-3}\}$ while, under the latter model, it becomes $\Sigma'_q = \{\text{M-2, M-3}\}$. The latter decision problem is now not quite as definitive for Melba as is the former one, where there was greater diversity between selectability and rejectability.

Diversity is infinite if there exist options with nonzero selectability and zero rejectability. Such options are free options, since no cost independent of avoiding failure is incurred by adopting them (analogy: coasting saves fuel, but may or may not get you to your destination). Diversity is not a measure of performance; that is, if one decision maker has a more diverse selectability/rejectability pair than another, that is not an indication that it will perform better than the other. It does, however, provide a measure of the conflict experienced by the decision maker.

Tension

An alternative measure of the equivocation experienced by the decision maker is to formulate a function that permits a convenient comparison with the state of being neutral with respect to attitude. Although the diversity functional provides insight into the relationship between selectability and rejectability, it does not afford a convenient comparison with the case where the decision maker is neutral with respect to either selectability or rejectability. To develop such a measure, it is convenient to re-normalize the selectability and rejectability functions. Consider first the case where p_S and p_R are mass functions and U is finite. Let

$$\mathbf{p}_S = [p_S(u_1), \ldots, p_S(u_n)],$$
$$\mathbf{p}_R = [p_R(u_1), \ldots, p_R(u_n)]$$

be selectability and rejectability row vectors, and let $\mu = [\frac{1}{n}, \ldots, \frac{1}{n}]$ denote the uniform mass function row vector, where n is the cardinality of U. Although these vectors are unit length under the L_1 norm, they are not of unit length under the L_2 norm. It will be convenient to normalize these vectors with respect to L_2. Let $|\mathbf{p}_S| = \sqrt{\mathbf{p}_S \mathbf{p}_S^{\mathsf{T}}}$, with similar definitions for $|\mathbf{p}_R|$ and $|\mu|$. The L_2 normalized mass function vectors will be denoted by $\tilde{\mathbf{p}}_S = \frac{\mathbf{p}_S}{|\mathbf{p}_S|}$ and similarly for $\tilde{\mathbf{p}}_R$ and μ.

We may express the similarity between p_S and p_R through the inner product of the corresponding unit vectors, yielding the expression $\tilde{\mathbf{p}}_S \tilde{\mathbf{p}}_R^{\mathsf{T}}$. This quantity will be unity when $p_S \equiv p_R$, and will decrease as the two mass functions tend to become orthogonal, thus capturing some of the properties we desire to model. If we normalize by the product of the projections of $\mathbf{p}_S$ and $\mathbf{p}_R$ onto the uniform distribution, we tend to scale up the inner product as the mass function vectors become distanced from the uniform distribution.

Definition 5.12
The **tension functional** is

$$T(p_S \| p_R) = \frac{\tilde{\mathbf{p}}_S \tilde{\mathbf{p}}_R^{\mathsf{T}}}{\tilde{\mathbf{p}}_S \tilde{\mu}^{\mathsf{T}} \tilde{\mathbf{p}}_R \tilde{\mu}^{\mathsf{T}}},$$

which simplifies into the convenient form:

$$T(p_S \| p_R) = n \mathbf{p}_S \mathbf{p}_R^{\mathsf{T}} = n \sum_{i=1}^{n} p_S(u_i) p_R(u_i).$$

□

When U is continuous, we consider the case where U is an interval, that is, $U = [a, b]$. A similar analysis reveals that the tension functional becomes

$$T(p_S \| p_R) = (b - a) \int_a^q p_S(u) p_R(u) du.$$

Theorem 5.2

For $U = \{u_1, \ldots, u_n\}$,

$$0 \leq T(p_S \| p_R) \leq n \min\{\max_i\{p_S(u_i)\}, \max_i\{p_R(u_i)\}\}.$$

If $p_S(u_i) = p_R(u_i)$ for $i = 1, \ldots, n$, then

$$1 \leq T(p_S \| p_R) \leq n \max_i\{p_S(u_i)\}.$$

For $U = [a, b]$,

$$0 \leq T(p_S \| p_R) \leq (b - a) \min\{\max_{u \in U}\{p_S(u)\}, \max_{u \in U}\{p_R(u)\}\}.$$

If $p_S(u) = p_R(u)$ for all $u \in U$, then

$$1 \leq T(p_S \| p_R) \leq (b - a) \max_{u \in U}\{p_S(u)\}.$$

If either p_S or p_R is uniform, then

$$T(p_S \| p_R) = 1.$$

PROOF

Clearly, $T(p_S \| p_R)$ is always non-negative. For $U = \{u_1, \ldots, u_n\}$,

$$n \sum_{i=1}^n p_S(u_i) p_R(u_i) \leq n \max_i\{p_S(u_i)\} \sum_{i=1}^n p_R(u_i) = n \max_i\{p_S(u_i)\}.$$

By a similar argument, $n \sum_{i=1}^n p_S(u_i) p_R(u_i) \leq n \max_i\{p_S(u_i)\}$. Consequently,

$$T(p_S \| p_R) \leq n \min\{\max_i\{p_S(u_i)\}, \max_i\{p_R(u_i)\}\}.$$

If $p_S(u_i) = p_R(u_i)$ for $i = 1, \ldots, n$, then, by Hölder's inequality,

$$\left(\sum_{i=1}^n \frac{1}{n} p_S(u_i) \right)^2 \leq \sum_{i=1}^n \frac{1}{n^2} \sum_{i=1}^n p_S^2(u_i).$$

Simplifying,

$$\frac{1}{n^2} \leq \frac{1}{n} \sum_{i=1}^n p_S^2(u_i),$$

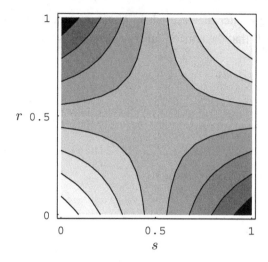

Figure 5.6: The contour plot of the tension functional for a two-dimensional decision problem.

or, upon rearranging,

$$1 \leq n \sum_{i=1}^{n} p_S^2(u_i) = T(p_S \| p_S) \leq n \max_i \{p_S(u_i)\}.$$

Suppose p_S is uniform. Then

$$n \sum_{i=1}^{n} \frac{1}{n} p_R(u_i) = \frac{n}{n} \sum_{i=1}^{n} p_R(u_i) = 1,$$

and similarly if p_R is uniform.

The proof for $U = [a, b]$ follows by a similar argument. □

If the rejectability function is uniform, then the decision maker is rejectability-neutral. If the selectability is uniform, then the decision maker is selectability-neutral. If $T(p_S \| p_R) > 1$, the decision maker is in a state of conflict, and if $T(p_S \| p_R) < 1$, the decision maker is in a state of contentment. When $n = 2$, and with selectability and rejectability mass vectors of the form $\mathbf{p}_S = [s, 1 - s]$ and $\mathbf{p}_R = [r, 1 - r]$, the tension function is

$$T(p_S, p_R) = 2[sr + (1 - s)(1 - r)].$$

Figure 5.6 illustrates tension for this case, with lighter shading indicating increased tension. Comparing this figure with Figures 5.3 and 5.4 illustrates that the tension function is consistent with the attitudinal properties of the decision maker, in that high-tension regions correspond to dispositions of dubiety or ambivalence, and low-tension regions correspond to dispositions of relief or gratification. Furthermore, when either s or r is close to 0.5 (resulting in the uniform distribution), the tension is near unity, indicating that the decision maker's attitude is neutral.

As an example of tension, consider Example 4.1, where Lucy is choosing a car. Referring to the selectability and rejectability values provided in Table 4.2, the tension is easily computed to be $T(p_S \| p_R) = 1.09$, which is somewhat less than the upper bound of $5 \times 0.267 = 1.335$, indicating that she is slightly conflicted in her decision.

A decision maker operating in the mode of gratification is well-tuned to its task. This is the situation when decisions that possess high selectability also possess low rejectability. Such a decision maker should be expected to achieve its goals with ease. A conservation-neutral decision maker will function much as would a conventional Bayesian decision maker. If it is failure neutral, it will function much like a minimax decision maker. If the decision maker is both conservation neutral and failure neutral, it is completely indifferent to the outcome, and there is little point in even attempting to make a decision other than randomly.

Diversity and tension are related concepts, but there are some significant differences. For example, comparing Figures 5.5 and 5.6 indicates that, along the line $s = r$, diversity is constant (zero), while tension reaches a maximum of $T(p_S, p_R) = 2$ at $s = r = 0$ and $s = r = 1$, but dips to a saddle point of $T(p_S, p_R) = 1$ at $s = r = \frac{1}{2}$. Diversity and tension, however, are *not* performance measures. They are simply figures of merit that evaluate the structure of the selectability and rejectability functions and serve to some extent as measures of the decision maker's functionality.

5.4   Quasi-invariance

As they are generally applied in decision making, utilities are used to make intra-utility comparisons; that is, they provide an ordering of preference among the options available to the decision maker. One of the key properties of von Neumann–Morgenstern utility theory is that the solution to a decision problem should be independent of the units (scale and origin) in which the utility is expressed. This is the **principle of invariance**, which holds that the solution should not change if utilities are scaled by positive affine transformations. Difficulties may arise, however, if multiple utilities are employed and the decision requires inter-utility comparisons; that is, comparing the value of one utility function with the value of another utility function. Such comparisons are generally not invariant to changes in scale or origin, and solutions may be affected by positive affine transformations.

The issue of inter-utility comparison, also termed interpersonal comparisons of utility, arises in traditional game theory, particularly in the theory of bargaining (Nash, 1950), and in social choice theory (Harsanyi, 1977). In bargaining, as well as in game theory generally, solutions are desired that are invariant to changes of scale and origin of the utility functions involved. Consequently, interpersonal comparisons are undesirable for such applications. Interpersonal comparisons are used extensively, however, in social choice theory. To make such comparisons meaningful, it is essential that the utility functions have the same origin and the same utility unit (Harsanyi, 1977).

The comparisons of the type required by praxeic utility theory are similar in some respects to the interpersonal comparisons made in social choice theory but are quite different in other respects. In the social choice context, a decision maker must assess the utilities that another separate decision maker would derive from various social situations and compare these with the utilities that the decision maker would derive from the same situations. For example, let X and Y be two decision makers, and suppose X's utility function is f_X and X's assessment of Y's utility function is $\hat{f}_Y$. If f_X and $\hat{f}_Y$ have the same scale and unit, then X may make assessments such as: "If $\hat{f}_Y(u) > f_X(u)$, then u would be of more benefit to Y than it would be to me." In the praxeic utility context, however, X must compare different attributes of each proposition by addressing questions such as: "If $p_S(u) > p_R(u)$, then I will derive more avoidance of failure than resource consumption by not rejecting u." These two types of comparisons are different in that the former involves utility comparisons involving different decision makers, whereas the latter involves comparisons involving different attributes of the decision. The former requires a decision maker to evaluate how another decision maker feels relative to itself, while the latter requires a decision maker to evaluate how it feels regarding different criteria. In the former case, scale and origin are issues because different decision makers may be operating with different units. In the latter case, scale and origin are issues because the decision maker may conceivably attach different units to the attributes of failure avoidance and resource consumption.

In one obvious sense, p_S and p_R are comparable since the special structure of p_S and p_R as mass functions automatically provides them each with a unit mass to apportion among the elements of U. To ascertain whether this structure is adequate to guarantee meaningful comparisons between mass functions, we must make a careful study of how these utilities behave under positive affine transformations. Let p be a mass function over an n-dimensional option space and let us transform this mass function using a positive affine transformation such that the transformed utility is also a mass function. Thus, we are restricted to transformations of the form

$$\pi(u) = \epsilon p(u) + \delta$$

with $\epsilon > 0$, subject to the constraints that $\pi(u) \geq 0$, $u \in U$, and $\sum_{u \in U} \pi(u) = 1$.

Definition 5.13

A transformation, T, of a mass function is said to be **isomassive** if it preserves the unit mass characteristics of the function; that is, if $T[p(u)] \geq 0$ and $\sum_{u \in U} T[p(u)] = 1$.

□

Theorem 5.3

Let p be a mass function over a space U consisting of n elements. The transformation

$$T_\epsilon[p(u)] = \epsilon p(u) + \frac{1 - \epsilon}{n}, \tag{5.4}$$

where $0 \le \epsilon \le 1$, is an affine isomassive transformation such that, if $\pi = T_\epsilon[p]$, then

$$p(u) \le \frac{1}{n} \text{ implies } p(u) \le \pi(u) \le \frac{1}{n}, \tag{5.5}$$

$$p(u) \ge \frac{1}{n} \text{ implies } p(u) \ge \pi(u) \ge \frac{1}{n}. \tag{5.6}$$

If $\epsilon > 1 - np(u^0)$, where

$$u^0 = \arg\min_{u \in U}\{p(u)\}, \tag{5.7}$$

then the transform

$$T_\epsilon^{-1}[p(u)] = \frac{1}{\epsilon}\left[p(u) - \frac{1-\epsilon}{n}\right] \tag{5.8}$$

is an affine isomassive transformation such that, if $\pi = T_\epsilon^{-1}[p]$, then π is a mass function such that

$$p(u) \le \frac{1}{n} \text{ implies } \pi(u) \le p(u), \tag{5.9}$$

$$p(u) \ge \frac{1}{n} \text{ implies } \pi(u) \ge p(u). \tag{5.10}$$

PROOF

Let $\pi(u) = \epsilon p(u) + \delta$. For π to qualify as a mass function, we must have $\pi(u) \ge 0$ for all $u \in U$. Since it is possible that $p(u) = 0$ for some $u \in U$, in which case $\pi(u) = \delta$, δ must be non-negative. We must also have $\sum_{u \in U} \pi(u) = 1$, or

$$\sum_{u \in U} [\epsilon p(u) + \delta] = \epsilon + n\delta = 1,$$

which means that $\delta = \frac{1-\epsilon}{n}$. The condition $\delta \ge 0$ thus implies $\epsilon \le 1$. Finally, since the transformation must be positive, ϵ must be non-negative, thus $0 \le \epsilon \le 1$.

To establish the ordering of p and π, suppose that $p(u) \le \frac{1}{n}$. Then

$$(1 - \epsilon)p(u) \le \frac{1-\epsilon}{n}$$

or, upon rearranging,

$$p(u) \le \epsilon p(u) + \frac{1-\epsilon}{n} = \pi(u).$$

Furthermore,

$$\pi(u) = \epsilon p(u) + \frac{1}{n}(1 - \epsilon)$$

$$= \epsilon\left(p(u) - \frac{1}{n}\right) + \frac{1}{n}$$

$$\le \frac{1}{n},$$

since $p(u) \leq \frac{1}{n}$. Thus, $p(u) \leq \pi(u) \leq \frac{1}{n}$. By a similar argument, it is easily established that, if $p(u) \geq \frac{1}{n}$, then $p(u) \geq \pi(u) \geq \frac{1}{n}$.

Also, it is immediately obvious that, if $\epsilon > 1 - np(u^0)$ and $\pi(u) = T_\epsilon^{-1}[p(u)]$, then $\pi(u) \geq 0$ for all $u \in U$ and $\sum_{u \in U} \pi(u) = 1$, so $\pi(u)$ is a mass function. Then $T_\epsilon^{-1}[\cdot]$ is the inverse transform of $T_\epsilon[\cdot]$, and $p(u) = T_\epsilon[\pi(u)]$, so (5.9) and (5.10) hold. $\square$

Observe that T_ϵ and its inverse transform, T_ϵ^{-1}, are order preserving so that, if $p(u_1) \leq p(u_2)$, then $T[p(u_1)] \leq T[p(u_2)]$. Note also, that the requirement that the transformation of a mass function must also be a mass function imposes an additional constraint that eliminates one of the degrees of freedom, so there is only one free parameter in the transformation, namely, ϵ. Consequently, there is no clear distinction between scale and origin for this class of utilities. Since there are no clear analogs to the notions of scale and origin for isomassive transformations, the notions do not apply, and we must look for an alternative criterion for meaningful comparisons using this family of utility functions.

To understand the effect of the transformation T_ϵ, we apply the notion of entropy.

Theorem 5.4

Let p and π be two mass functions defined on a space U consisting of n elements such that the relationships

$$p(u) \geq \frac{1}{n} \text{ implies } 1 > p(u) \geq \pi(u) \geq \frac{1}{n}, \tag{5.11}$$

$$p(u) = \frac{1}{n} \text{ implies } p(u) = \pi(u) = \frac{1}{n}, \tag{5.12}$$

$$p(u) \leq \frac{1}{n} \text{ implies } 0 < p(u) \leq \pi(u) \leq \frac{1}{n} \tag{5.13}$$

hold. Then $H(\pi) \geq H(p)$.

PROOF

Let $\Delta(u) = \pi(u) - p(u)$ and define

$$U_+ = \left\{ u \in U : p(u) > \frac{1}{n} \right\},$$

$$U_0 = \left\{ u \in U : p(u) = \frac{1}{n} \right\},$$

$$U_- = \left\{ u \in U : p(u) < \frac{1}{n} \right\}.$$

Then

$$H(\pi) - H(p) = \sum_U [p(u) + \Delta(u)] \log_2 \frac{1}{\pi(u)} + \sum_u p(u) \log_2 p(u)$$

$$= \sum_U p(u) \log_2 \frac{p(u)}{\pi(u)} + \sum_U \Delta(u) \log_2 \frac{1}{\pi(u)}$$

$$= D(p\|\pi) + \sum_{U_+} \Delta(u) \log_2 \frac{1}{\pi(u)}$$

$$+ \sum_{U_0} \Delta(u) \log_2 n + \sum_{U_-} \Delta(u) \log_2 \frac{1}{\pi(u)}.$$

For $u \in U_+$, (5.11) implies that $\Delta(u) \le 0$ and

$$\log_2 \frac{1}{\pi(u)} \le \log_2 n,$$

so

$$\Delta(u) \log_2 \frac{1}{\pi(u)} \ge \Delta(u) \log_2 n.$$

Also, for $u \in U_-$, $\Delta(u) \ge 0$ and

$$\log_2 \frac{1}{\pi(u)} \ge \log_2 n,$$

so

$$\Delta(u) \log_2 \frac{1}{\pi(u)} \ge \Delta(u) \log_2 n.$$

Thus,

$$H(\pi) - H(p) \ge D(p\|\pi) + \sum_U \Delta(u) \log_2 n = D(p\|\pi) \ge 0,$$

since $\sum_U \Delta(u) = 0$.   □

From Theorems 5.3 and 5.4 we may conclude that T_ϵ is an entropy-increasing transformation and T_ϵ^{-1} is an entropy-decreasing transformation. The effect of applying T_ϵ is to decrease the dynamic range of the mass function to make it flatter, or more uniform. The flatter a distribution becomes, the higher its entropy. The more peaked it becomes, the lower its entropy. Note that, if p is uniform, then it is a fixed point of the transformation, that is, $p = T_\epsilon[p]$ for all values of ϵ. Any distribution that places all of its mass on one point has zero entropy. It is not possible to transform an arbitrary mass function to a zero-entropy mass function by means of an affine isomassive transformation. However, let us define the minimum-entropy transform as the transform $T_{\epsilon^0}^{-1}$, where $\epsilon^0 = 1 - p(u^0)$, and u^0 is given by (5.7). Under this transformation, the transformed mass function of at least one element of U, namely, u^0, will have zero mass.

Thus, isomassive transformations serve only to modulate the entropy of the mass function without changing the intra-probability orderings. T_ϵ serves to make the hills of the mass function less steep and the valleys less deep as ϵ becomes smaller. T_ϵ^{-1} serves to make the hills of the mass function steeper and the valleys deeper as ϵ becomes smaller.

To evaluate the comparability of p_S and p_R we set $q = 1$, the nominal value of boldness that assigns equal weight to selectability and rejectability.

Definition 5.14

An option $u \in U$ is **invariant with respect to disposition** if arbitrary isomassive transformations of p_S and p_R do not change the dispositional classification of u.   □

Definition 5.15

Let p'_S be an arbitrary isomassive transformation of p_S and p'_R be an arbitrary isomassive transformation of p_R. An option $u \in U$ is **invariant with respect to ordering** if $p_S(u) \geq p_R(u)$ implies $p'_S(u) \geq p'_R(u)$ and $p_S(u) \leq p_R(u)$ implies $p'_S(u) \leq p'_R(u)$.   □

Definition 5.16

An option that is invariant with respect to disposition and is invariant with respect to ordering for the modes of relief and gratification is said to be **quasi-invariant**.   □

Let U be a finite set of options with selectability and rejectability functions p_S and p_R, respectively.

Theorem 5.5

Every $u \in U$ is quasi-invariant.

PROOF

Let $p'_S = T_{\epsilon_1}[p_S]$ be any isomassive transform of p_S, and let $p'_R = T_{\epsilon_2}[p_R]$ be any isomassive transform of p_R. If $p_R(u) > \frac{1}{n}$ and $p_S(u) < \frac{1}{n}$ (relief), then by Theorem 5.3, $p'_S(u) < p'_R(u)$. Also, if $p_R(u) < \frac{1}{n}$ and $p_S(u) > \frac{1}{n}$ (gratification), then $p'_R(u) < p'_S(u)$. Now suppose $p_R(u) > \frac{1}{n}$ and $p_S(u) > \frac{1}{n}$ (ambivalence) or $p_R(u) < \frac{1}{n}$ and $p_S(u) < \frac{1}{n}$ (dubiety). By Theorem 5.3, isomassive transformations preserve these dispositional modes, since $p'_R(u) > \frac{1}{n}$ and $p'_S(u) > \frac{1}{n}$ or $p'_R(u) < \frac{1}{n}$ and $p'_S(u) < \frac{1}{n}$. Inter-utility ordering is not preserved for these dispositions, however, since it is not necessarily true that, for example, $p_S(u) > p_R(u)$ implies $p'_S(u) > p'_R(u)$. Thus, comparisons of p_S and p_R are invariant with respect to disposition for all dispositional modes and are invariant with respect to ordering for the modes of relief and gratification.   □

Quasi-invariance characterizes the properties of satisficing decisions with boldness of one. Invariance with respect to disposition tells us that isomassive transformations cannot alter the basic attitudinal properties of a decision problem. If a decision is gratifying under one set of selectability and rejectability pairs, it will remain so under any isomassive transformations of the pair; similarly for the other modes.

With respect to the modes of contentment (gratification and relief), the theorem also tells us that these modes are stable, in that isomassive transformations will not alter the ordering of selectability and rejectability and hence will not alter the decision (at least with unity boldness). To illustrate, let p_S' be an isomassive transform of p_S, and suppose u is such that $p_R(u) > \frac{1}{n}$ and $p_S(u) < \frac{1}{n}$. Then $p_S'(u) < p_R(u)$ for all admissible ϵ. In other words, an option that was formerly very unattractive cannot be made attractive by an isomassive transformation. Similarly, if $p_S(u) > \frac{1}{n}$ and $p_R(u) < \frac{1}{n}$, then $p_S'(u) > p_R(u)$, meaning that a highly attractive option cannot be made unattractive by an isomassive transformation.

The dispositional modes of ambivalence and dubiety are the modes of conflict. In these modes, decision making is the most difficult. Although these modes cannot be altered by isomassive transformations, the selectability–rejectability orderings within the regions U_A and U_D are not invariant. It is not surprising that these modes would be sensitive to the precise numerical values of the selectability and rejectability functions. They are necessarily unstable modes.

6 Community

Unless we make ourselves hermits, we shall necessarily influence each other's opinions; so that the problem becomes how to fix belief, not in the individual merely, but in the community.

Charles Sanders Peirce
The Fixation of Belief, Popular Science Monthly (1877)

Decision makers are usually not hermits and do not function in isolation from others. They are usually influenced by the opinions (i.e., preferences) of other decision makers, and their problem is one of how to select a course of action, not only for themselves, but for the community. If each individual were to possess a notion of rationality for the group as well as a notion of rationality for itself, it might be in a position to improve its behavior.

Group rationality, however, is not a logical consequence of rationality based on individual self-interest. Under substantive rationality, where maximization of individual satisfaction is the operative notion, group behavior is not usually optimized by optimizing each individual's behavior, as is done in conventional game theory. Unfortunately, those who put their final confidence in exclusive self-interest may ultimately function disjunctively, and perhaps illogically, when participating in collective decisions.

The point of departure for conventional game theory is games of pure conflict, with the prototype being constant-sum games, where one player's loss is another player's gain. For a constant-sum game, any notion of group-interest is vacuous; individual self-interest is the only appropriate motive. Non-constant-sum games represent non-purely conflictive situations, but though a vast theory of games has been developed, its main emphasis is on conflict. Schelling (1960) has developed what he calls a "reorientation" of game theory, wherein he attempts to counterbalance the notion of conflict with a game-theoretic notion of coordination. The antipodal extreme for a game of pure conflict is a game of pure coordination, where all players have coincident interests and desire to take actions that are simultaneously of self and mutual benefit. In contrast to pure conflict games where one player wins only if the others lose, in a pure coordination game either all players win or all lose. In general, games may involve both conflict and coordination; Schelling terms such games mixed-motive games.

Table 6.1: The payoff matrix for a zero-sum game with a coordination equilibrium

	X_2	
X_1	C_1	C_2
R_1	$(0, 0)$	$(0, 0)$
R_2	$(0, 0)$	$(1, -1)$

Although Schelling's attempt at reorientation is useful, it is not a fully adequate way to mitigate the notion of conflict, since his reorientation does not alter the fundamental structure of game theory as ultimately dependent on the principle of individual rationality. Lewis attempts to patch up the problem by introducing the notion of coordination equilibrium as a refinement of Nash equilibrium. Whereas a Nash equilibrium is a combination in which no one would be better off had he alone acted otherwise, Lewis (1969) strengthens this concept and defines a coordination equilibrium.

Definition 6.1

A **coordination equilibrium** is a combination in which no one decision maker would have been better off had *any one* decision maker alone acted otherwise, either himself or someone else (Lewis, 1969, p. 14). ☐

Coordination equilibria are common in situations of mixed opposition and coincidence of interest. In fact, even a zero-sum game can have a coordination equilibrium, as is evident from the example displayed in Table 6.1, taken from Lewis (1969), where (R_1, C_1) is a coordination equilibrium. Attempts to reorient game theory, while perhaps desirable for coordination scenarios, are not entirely successful; the issue of mitigating individual rationality is not resolved.

Rather than reorienting game theory to accommodate situations where coordination is a more natural operational descriptor of the game than is conflict, I propose to change the foundational cornerstone of the theory. My goal is to arrive at a notion that is completely neutral with respect to conflictive and coordinative descriptors and equally accommodates both of them. By so doing, we may unify the solution concept so that there will not be a need to seek different equilibria.

The price to be paid for complete conflict/coordination neutrality is the abandonment of the rigid principle of optimization, which I propose to replace with the more pliable notion of satisficing. Satisficing admits degrees of fulfillment, whereas optimization is absolute. Thus, while the assertion "What is best for me and what is best for you is also jointly best for us together" may be nonsense, the statement "What is good enough for me and what is good enough for you is also jointly good enough for us together" may be perfectly sensible, so long as the notions of what it means to be good enough are not inflexible. Satisficing grants room for compromise. There is the opportunity for one or both parties to relax their standards of individual performance in the interest of the good of the dyad.

This chapter presents a theory of multi-agent decision making that is based on the new concept of satisficing presented in Chapter 3. The result is a new theory of games, called satisficing games (Stirling and Goodrich, 1999a). The development proceeds as follows: first, a way to express both group and individual preferences is defined in a way that accounts for the interdependencies that exist between decision makers. Next, a multi-agent version of praxeic utility theory is developed. Finally, the consequences of this model of group and individual behavior are investigated.

6.1   Joint and individual options

We have seen that, in addition to the superlative paradigm, a comparative paradigm for decision making under conditions of intrinsic rationality is possible. Further, we have developed formalisms for such a paradigm based on the dichotomy between avoiding failure (selectability) and conserving resources (rejectability). Key features of this paradigm are: (a) it admits set-valued, rather than point-valued decisions, and (b) decisions are based on a notion of satisficing that has been formalized and made mathematically precise through the application of praxeic utility theory.

Thus far, we have applied praxeic utility theory to single decision makers, but there is no reason, as far as the theory is concerned, why it cannot be extended to the multi-agent case, with the option space consisting of option vectors where each element of the vector corresponds to a particular individual. In fact, in the development of the game in Example 3.4, we viewed Nature as at least a passive player of the game and formed the utilities accordingly.

Since they possess the mathematical structure of probabilities, selectability and rejectability can be extended naturally to the multivariate (i.e., multi-agent) case by defining joint selectability and rejectability measures which may then be used to evaluate a joint dichotomy to establish the notion of joint satisficing for the community. In addition, individual decision makers may establish individual notions of satisficing by computing their marginal selectability and rejectability values from the joint expressions. In this way, both group and individual preferences may be expressed.

Definition 6.2
An N**-member multi-agent system** (N a positive integer), denoted $\mathbf{X} = \{X_1, \ldots, X_N\}$, is a collection of N decision makers, each with its option space U_i, $i = 1, \ldots, N$.   □

6.2   Interdependency

Our ultimate goal is to construct joint selectability and joint rejectability functions from which we may identify the joint options that are satisficing from the point of view of the entire community. We may then extract the individual selectability and rejectability functions as the marginals of these joint functions, from which we may identify the options that are individually satisficing. To achieve this goal, however, we must first deal with a more fundamental issue: the interconnections between selectability and rejectability.

In our development of praxeic utility theory, we determined the selectability of options by examining them in the light of the support they lend to the avoidance of failure, without taking into consideration the conservation of resources. We also determined the rejectability of options by examining them strictly in terms of the resources they consume, without taking avoidance of failure into consideration. This separation of interests is possible when only a single decision maker is involved, but in an environment of multiple decision makers, it is not generally possible to determine selectability and rejectability in isolation from each other.

An act by any individual member of a multi-agent system has possible ramifications for the entire system. Some participants may be benefited by the act, some may be damaged, and some may be indifferent. Furthermore, although an individual may perform the act in its own interest or for the benefit of others or the entire system, the act is usually not implemented free of cost. Resources are expended, or risk is taken, or some other penalty or unpleasant consequence is incurred, perhaps by the individual whose act it is, perhaps by other participants, and perhaps by the entire system. Although these undesirable consequences may be defined independently from the benefits, the measures associated with benefits and costs cannot be specified independently of each other, due to the possibility of interaction. A critical aspect of modeling the behavior of such a system, therefore, is the means of representing the interdependence of both positive and negative consequences of all possible joint options that could be undertaken.

Example 6.1 Lucy and Ricky are going to buy the family car. Lucy gets to choose the make, either a Porsche (P) or a Honda (H), and Ricky gets to choose the number of seats, either two (T) or four (F). Selectability for them is defined in terms of style, and rejectability is measured in terms of the number of passengers that can be accommodated (the fewer the passengers, the greater the consumption and hence the lower the conservation).

Being praxeic utility enthusiasts, Lucy and Ricky wish to identify a set of satisficing joint options. To do so, they must compute their joint selectability and rejectability mass functions, denoted by

$$p_{S_L S_R}: \{P, H\} \times \{T, F\} \to [0, 1],$$
$$p_{R_L R_R}: \{P, H\} \times \{T, F\} \to [0, 1],$$

respectively.[1] These functions can be determined in isolation from each other only if issues of joint selectability are completely separated from issues of joint rejectability. This would mean, for example, that even if Lucy knew that that Ricky were committed to rejecting T, she would not change her assessment of the selectability of P (assuming she at least unconditionally highly favors P), even though the joint option (P, F) is impossible. If, however, such conditional knowledge would influence Lucy's assessment, then the joint selectability and joint rejectability cannot be specified in isolation.

This example illustrates that we must look to a more fundamental structure to capture the inter-relationships that may exist between joint selectability and joint rejectability.

6.2.1   Mixtures

A natural way to account for inter-relationships is to express joint selectability and joint rejectability as marginals of a more general measure, which we may call interdependency. This will require us to examine subsets of options by multiple players. To develop the notion of interdependence and to define the interdependence measure, we require a number of definitions.

Definition 6.3
A **mixture** is any subset of decision makers considered in terms of their interaction with each other, exclusively of possible interactions with other decision makers not in the subset. A **selectability mixture**, denoted $\mathcal{S} = S_{i_1} \cdots S_{i_k}$, is a mixture consisting of $X_{i_1}, \ldots, X_{i_k}$ that is considered exclusively from the point of view of avoidance of failure. The **joint selectability mixture** is the selectability mixture consisting of all decision makers in the multi-agent system, denoted $\mathbf{S} = S_1 \cdots S_N$. A **myopic** selectability mixture is a mixture of the form $\mathcal{S} = S_i$. It occurs when an individual decision maker views its avoidance of failure as though it were functioning with complete disregard for all other decision makers. A **rejectability mixture**, denoted $\mathcal{R} = R_{j_1} \cdots R_{j_\ell}$, is a mixture consisting of $X_{j_1}, \ldots, X_{j_\ell}$ and is considered exclusively from the point of view of resource conservation. The **joint rejectability mixture** is the rejectability mixture consisting of all decision makers in the system, denoted $\mathbf{R} = R_1 \cdots R_N$. A **myopic** rejectability mixture is a mixture of the form $\mathcal{R} = R_i$.

An **intermixture** is the concatenation of a selectability mixture and a rejectability mixture and is denoted $\mathcal{SR} = S_{i_1} \cdots S_{i_k} R_{j_1} \cdots R_{j_\ell}$. The **joint intermixture** is the concatenation of the joint selectability and joint rejectability mixtures and is denoted $\mathbf{SR} = S_1 \cdots S_N R_1 \cdots R_N$. A **myopic** intermixture is a mixture of the form $\mathcal{SR} = S_i R_i$. □

---

[1] In general, the product set notation notation $A \times B$, for any sets A and B, means the set of all ordered pairs (x, y), where $x \in A$ and $y \in B$.

Definition 6.4

Given arbitrary option spaces U_{i_j}, $j = 1, \ldots, m$, the **product option space**, denoted $U_{i_1} \times \cdots \times U_{i_m}$, is the set of all m-tuples $\mathbf{u} = (u_{i_1}, \ldots, u_{i_m})$ where $u_{i_j} \in U_{i_j}$, $j = 1, \ldots, m$. A **rectangle** in $U_{i_1} \times \cdots \times U_{i_m}$ is a subset of the form

$$F_{i_1} \times \cdots \times F_{i_m} = \{(u_{i_1}, \ldots, u_{i_m}): u_{i_j} \in F_{i_j}, j = 1, \ldots, m\}.$$

The **joint (selectability or rejectability) option space** is the product space $\mathbf{U} = U_1 \times \cdots \times U_N$. Elements of $\mathbf{U}$ are called **option vectors** or **joint options**. The **selectability subspace** associated with a selectability mixture $\mathcal{S} = S_{i_1} \cdots S_{i_k}$ is the product space $\mathbf{U}_{\mathcal{S}} = U_{i_1} \times \cdots \times U_{i_k}$. Elements of $\mathbf{U}_{\mathcal{S}}$ are called **selectability sub-vectors**. The **rejectability subspace** associated with a rejectability mixture $\mathcal{R} = R_{j_1} \cdots R_{j_\ell}$ is the product space $\mathbf{U}_{\mathcal{R}} = U_{j_1} \times \cdots \times U_{j_\ell}$. Elements of $\mathbf{U}_{\mathcal{R}}$ are called **rejectability sub-vectors**.

The **joint interaction space** is the product space $\mathbf{U} \times \mathbf{U}$. Elements of $\mathbf{U} \times \mathbf{U}$ are called **interaction vectors**. The **interaction subspace** associated with an intermixture $\mathcal{S}\mathcal{R} = S_{i_1} \cdots S_{i_k} R_{j_1} \cdots R_{j_\ell}$ is the product space $\mathbf{U}_{\mathcal{S}\mathcal{R}} = U_{i_1} \times \cdots \times U_{i_k} \times U_{j_1} \times \cdots \times U_{j_\ell}$, Elements of $\mathbf{U}_{\mathcal{S}\mathcal{R}}$ are called **interaction sub-vectors**.   □

Definition 6.5

A **selectability measure** $P_{\mathcal{S}}$ for the mixture $\mathcal{S} = S_{i_1} \cdots S_{i_k}$ is a normalized measure (i.e., $P_{\mathcal{S}}(\mathbf{U}_{\mathcal{S}}) = 1$) defined over the measurable subsets of $\mathbf{U}_{\mathcal{S}}$.[2] We will denote this measure as $P_{\mathcal{S}}$ or, alternatively, as $P_{S_{i_1} \cdots S_{i_k}}$. The **joint selectability measure** $P_{\mathbf{S}}$ is a selectability measure for the joint selectability mixture $\mathbf{S}$. A **rejectability measure** for the mixture $\mathcal{R}$ is a normalized measure $P_{\mathcal{R}} = P_{R_{j_1} \cdots R_{j_\ell}}$. The **joint rejectability measure** $P_{\mathbf{R}}$ is a rejectability measure for the joint rejectability mixture $\mathbf{R}$.

An **interdependence measure** $P_{\mathcal{S}\mathcal{R}}$ for the intermixture $\mathcal{S}\mathcal{R}$ is a normalized measure defined over the measurable subsets of $\mathbf{U}_{\mathcal{S}} \times \mathbf{U}_{\mathcal{R}}$. The **joint interdependence measure**, $P_{\mathbf{S}\mathbf{R}}$, is an interdependence measure for the joint intermixture $\mathbf{S}\mathbf{R}$. A **myopic interdependence measure** for a single agent is of the form $P_{S_i R_i}$, but, because of the requirement that single-agent selectability and rejectability be specified independently, a myopic interdependence measure must factor into the product of the myopic selectability and rejectability measures, that is, $P_{S_i R_i} = P_{S_i} P_{R_i}$ for $i = 1, 2, \ldots, N$.[3]   □

The joint interdependence measure provides a complete description of all individual and group relationships in terms of their positive and negative consequences.

---

[2] A measurable set is an element of a σ-field of sets (i.e., a collection of sets that is closed under complementation and countable unions). When the U_is are finite the σ-field will be taken as the power set of $\mathbf{U}_{\mathcal{S}}$, and when the U_is are a continuum, the σ-field is the Borel field over $\mathbf{U}_{\mathcal{S}}$. Normalized measures are probability measures, but to emphasize that the context here does not involve randomness, we use the term selectability and rejectability in lieu of probability.

[3] This situation is analogous to the probabilistic notion of independence, whereby the probability of independent events factors into the product of the probabilities of the events.

Let $\mathbf{V} \subset \mathbf{U}$ and $\mathbf{W} \subset \mathbf{U}$ be two sets of option vectors; that is, $\mathbf{V} \times \mathbf{W}$ is a rectangle in $\mathbf{U} \times \mathbf{U}$. Then $P_{SR}(\mathbf{V} \times \mathbf{W})$ is a representation of the failure avoidance associated with $\mathbf{V}$ and the resource conservation associated with $\mathbf{W}$ when the two option vector sets are viewed simultaneously. In other words, $P_{SR}(\mathbf{V} \times \mathbf{W})$ characterizes the disposition of the decision-making system with respect to selecting $\mathbf{V}$ (in the interest of avoiding failure) and rejecting $\mathbf{W}$ (in the interest of conservation). Particularly when $\mathbf{V} \cap \mathbf{W} \neq \emptyset$, it may appear contradictory to consider simultaneously rejecting and selecting options. It is important to remember, however, that considerations of selection and rejection involve two different criteria. It is no contradiction to consider selecting, in the interest of achieving a goal, an option that one would wish to reject for unrelated reasons, nor is it a contradiction to consider rejecting, because of some undesirable consequences, an option one would otherwise wish to select. Evaluating such tradeoffs is an essential part of decision making, and the interdependence measure provides a means of quantifying all issues relevant to this tradeoff.

6.2.2   Conditioning

It may appear that the construction of the interdependence measure will be a formidable task. Fortunately, however, the fact that it is a probability measure provides a greatly simplified means of constructing it. First, we require some additional definitions.

Definition 6.6
Given an intermixture $\mathcal{SR} = S_{i_1} \cdots S_{i_k} R_{j_1} \cdots R_{j_\ell}$, a **sub-intermixture** of $\mathcal{SR}$ is an intermixture formed by concatenating subsets of $\mathcal{S}$ and $\mathcal{R}$:

$$\mathcal{S}_1 \mathcal{R}_1 = S_{i_{p_1}} \cdots S_{i_{p_q}} R_{j_{r_1}} \cdots R_{j_{r_s}},$$

where $\{i_{p_1}, \ldots, i_{p_q}\} \subset \{i_1, \ldots, i_k\}$ and $\{j_{r_1}, \ldots, j_{r_s}\} \subset \{j_1, \ldots, j_\ell\}$. We shall use the notation $\mathcal{S}_1 \mathcal{R}_1 \subset \mathcal{SR}$ to indicate that $\mathcal{S}_1 \mathcal{R}_1$ is a sub-intermixture of $\mathcal{SR}$.

The $\mathcal{SR}$-**complementary sub-intermixture** associated with a sub-intermixture $\mathcal{S}_1 \mathcal{R}_1$ of an intermixture $\mathcal{SR}$, denoted $\mathcal{SR} \backslash \mathcal{S}_1 \mathcal{R}_1$, is an intermixture created by concatenating the selectability and rejectability mixtures formed by the relative complements of $\mathcal{S}$ and $\mathcal{R}$; that is, $\mathcal{SR} \backslash \mathcal{S}_1 \mathcal{R}_1 = (\mathcal{S} \backslash \mathcal{S}_1)(\mathcal{R} \backslash \mathcal{R}_1)$. Clearly, $\mathcal{SR} \backslash \mathcal{S}_1 \mathcal{R}_1 \subset \mathcal{SR}$. $\mathcal{SR}$ is the union of $\mathcal{SR} \backslash \mathcal{S}_1 \mathcal{R}_1$ and $\mathcal{S}_1 \mathcal{R}_1$, denoted $\mathcal{SR} = \mathcal{SR} \backslash \mathcal{S}_1 \mathcal{R}_1 \cup \mathcal{S}_1 \mathcal{R}_1$.   □

We require that interdependence measures be consistent, that is, for any sub-intermixture $\mathcal{S}_1 \mathcal{R}_1$ of an intermixture $\mathcal{SR}$ we require

$$P_{\mathcal{S}_1 \mathcal{R}_1}(\mathbf{F}) = P_{\mathcal{SR}}(\mathbf{F} \times \mathbf{U}_{\mathcal{SR} \backslash \mathcal{S}_1 \mathcal{R}_1}),$$
$$P_{\mathcal{SR} \backslash \mathcal{S}_1 \mathcal{R}_1}(\mathbf{G}) = P_{\mathcal{SR}}(\mathbf{U}_{\mathcal{S}_1 \mathcal{R}_1} \times \mathbf{G})$$

for all rectangles $\mathbf{F} \times \mathbf{G} \subset \mathbf{U}_{\mathcal{S}_1 \mathcal{R}_1} \times \mathbf{U}_{\mathcal{SR} \backslash \mathcal{S}_1 \mathcal{R}_1}$. This requirement means that the selectability and rejectability measures are consistent with the interdependence measure, that is,

$$P_{S_{i_1}\cdots S_{i_k}}(V_{i_1} \times \cdots \times V_{i_k}) = P_{S_{i_1}\cdots S_{i_k} R_{j_1}\cdots R_{j_\ell}}(V_{i_1} \times \cdots \times V_{i_k} \times U_{j_1} \times \cdots \times U_{j_\ell})$$

and

$$P_{R_{j_1}\cdots R_{j_\ell}}(W_{j_1} \times \cdots \times W_{j_\ell}) = P_{S_{i_1}\cdots S_{i_k} R_{j_1}\cdots R_{j_\ell}}(U_{i_1} \times \cdots \times U_{i_k} \times W_{j_1} \times \cdots \times W_{j_\ell}).$$

In particular, the joint selectability and joint rejectability measures are consistent with the joint interdependence measure:

$$P_{\mathbf{S}}(\mathbf{V}) = P_{\mathbf{SR}}(\mathbf{V} \times \mathbf{U}), \tag{6.1}$$

$$P_{\mathbf{R}}(\mathbf{W}) = P_{\mathbf{SR}}(\mathbf{U} \times \mathbf{W}). \tag{6.2}$$

The primary vehicle for relating intermixtures and sub-intermixtures is through the use of transition functions (as defined, for example, by Neveu (1965). We may construct the joint interdependence measure by partitioning the system into intermixtures and then relating the intermixtures through transition interdependence functions.

Definition 6.7
Let $\mathcal{SR}$ be an intermixture and let $\mathcal{S}_1 \mathcal{R}_1$ be a sub-intermixture of $\mathcal{SR}$. A **transition interdependence function**, denoted $P_{\mathcal{SR}\backslash\mathcal{S}_1\mathcal{R}_1|\mathcal{S}_1\mathcal{R}_1}$, is a mapping such that:
(a) for every $\mathbf{u} \in \mathbf{U}_{\mathcal{S}_1\mathcal{R}_1}$, $P_{\mathcal{SR}\backslash\mathcal{S}_1\mathcal{R}_1|\mathcal{S}_1\mathcal{R}_1}(\cdot|\mathbf{u})$ is a normalized measure over $\mathbf{U}_{\mathcal{SR}\backslash\mathcal{S}_1\mathcal{R}_1}$; and
(b) for every measurable set $\mathbf{F} \subset \mathbf{U}_{\mathcal{SR}\backslash\mathcal{S}_1\mathcal{R}_1}$, $P_{\mathcal{SR}\backslash\mathcal{S}_1\mathcal{R}_1|\mathcal{S}_1\mathcal{R}_1}(\mathbf{F}|\cdot)$ is a point-valued function on $\mathbf{U}_{\mathcal{S}_1\mathcal{R}_1}$.
Let us assume that all transition interdependence functions are consistent with interdependence measures. If $\mathcal{SR}$ is an arbitrary intermixture with sub-intermixture $\mathcal{S}_1\mathcal{R}_1$,

$$P_{\mathcal{SR}}(\mathbf{F} \times \mathbf{G}) = \int_{\mathbf{F}} P_{\mathcal{SR}\backslash\mathcal{S}_1\mathcal{R}_1|\mathcal{S}_1\mathcal{R}_1}(\mathbf{G}|\mathbf{f}) P_{\mathcal{S}_1\mathcal{R}_1}(d\mathbf{f}) \tag{6.3}$$

for all measurable $\mathbf{F} \subset \mathbf{U}_{\mathcal{S}_1\mathcal{R}_1}$ and all measurable $\mathbf{G} \subset \mathbf{U}_{\mathcal{SR}\backslash\mathcal{S}_1\mathcal{R}_1}$. □

Equation (6.3) provides the essential mechanism for relating behavior considered in isolation to behavior considered in a group and serves as an important building block for constructing the joint interdependence measure from local components. To see how this is accomplished, let $\mathcal{S}_1\mathcal{R}_1$ be an intermixture in **SR** and let $\mathcal{S}_2\mathcal{R}_2$ be a sub-intermixture in $\mathcal{S}_1\mathcal{R}_1$. Then

$$P_{\mathbf{SR}}(\mathbf{F}) = \int_{\mathbf{F}_{\mathcal{S}_1\mathcal{R}_1}} P_{\mathbf{SR}\backslash\mathcal{S}_1\mathcal{R}_1|\mathcal{S}_1\mathcal{R}_1}(\mathbf{F}_{\mathbf{SR}\backslash\mathcal{S}_1\mathcal{R}_1}|\mathbf{f}_{\mathcal{S}_1\mathcal{R}_1}) P_{\mathcal{S}_1\mathcal{R}_1}(d\mathbf{f}_{\mathcal{S}_1\mathcal{R}_1}),$$

where $\mathbf{F} = \mathbf{F}_{\mathbf{SR}\backslash\mathcal{S}_1\mathcal{R}_1} \cup \mathbf{F}_{\mathcal{S}_1\mathcal{R}_1}$. Using the fact that $P_{\mathcal{S}_1\mathcal{R}_1}$ can also be expressed in terms of a transition interdependence function between S_2R_2 and $S_1R_1 \setminus S_2R_2$, we obtain

$$P_{\mathbf{SR}}(\mathbf{F}) = \int_{\mathbf{F}_{S_1R_1}} P_{\mathbf{SR}\setminus S_1R_1|S_1R_1}(\mathbf{F}_{\mathbf{SR}\setminus S_1R_1}|\mathbf{f}_{S_1R_1})$$
$$\cdot P_{S_1R_1\setminus S_2R_2|S_2R_2}(d\mathbf{f}_{S_1R_1\setminus S_2R_2}|\mathbf{f}_{S_2R_2}) P_{S_2R_2}(d\mathbf{f}_{S_2R_2}), \qquad (6.4)$$

where $\mathbf{F}_{S_1R_1} = \mathbf{F}_{S_1R_1\setminus S_2R_2} \cup \mathbf{F}_{S_2R_2}$.

The basic concept of interdependency is cast in measure theory, but, as is the case with probability theory in general, for applications it is more useful to work with mass functions (for discrete option spaces) or density functions (for continuous option spaces) rather than directly with interdependence measures.

Definition 6.8

Let $\mathbf{U} = U_1 \times \cdots \times U_N$, where each U_i contains countably many elements. Let $\mathcal{F}_i = 2^{U_i}$, the power set of U_i and let P_{SR} be an interdependence measure for the intermixture SR. An **interdependence mass function** is a mass function, p_{SR}, defined by

$$p_{SR}(\mathbf{v}) = P_{SR}(\{\mathbf{v}\}),$$

where $\mathbf{v} \in \mathbf{U}_{SR}$ and $\{\mathbf{v}\}$ is the singleton set consisting of $\mathbf{v}$. $\qquad\square$

To keep notation as clear as possible, for a selectability mixture $S = S_{i_1} \cdots S_{i_k}$ we write $p_S(\mathbf{s})$ or, equivalently, $p_S(s_{i_1}, \ldots, s_{i_k})$, or yet again equivalently, $p_{S_{i_1}\cdots S_{i_k}}(s_{i_1}, \ldots, s_{i_k})$, for $\mathbf{s} = \{s_{i_1}, \ldots, s_{i_k}\}$, where $\mathbf{s} \in \mathbf{U}_S$ is a selectability sub-vector. Similarly, for a rejectability mixture $\mathcal{R} = R_{j_1} \cdots R_{j_\ell}$ we write $p_{\mathcal{R}}(\mathbf{r})$ or, equivalently, $p_{\mathcal{R}}(r_{j_1}, \ldots, r_{j_\ell})$, or yet again equivalently, $p_{R_{j_1}\cdots R_{j_\ell}}(r_{j_1}, \ldots, r_{j_\ell})$, for $\mathbf{r} = \{r_{j_1}, \ldots, r_{j_\ell}\}$, where $\mathbf{r} \in \mathbf{U}_{\mathcal{R}}$ is a rejectability sub-vector. Finally, for an intermixture $SR = S_{i_1} \cdots S_{i_k} R_{j_1} \cdots R_{j_\ell}$ we write $p_{SR}(\mathbf{s}; \mathbf{r})$ or, equivalently, $p_{SR}(s_{i_1}, \ldots, s_{i_k}; r_{j_1}, \ldots, r_{j_\ell})$, or yet again equivalently, $p_{S_{i_1}\cdots S_{i_k} R_{j_1}\cdots R_{j_\ell}}(s_{i_1}, \ldots, s_{i_k}; r_{j_1}, \ldots, r_{j_\ell})$ for sub-vectors $\mathbf{s} = \{s_{i_1}, \ldots, s_{i_k}\}$ and $\mathbf{r} = \{r_{j_1}, \ldots, r_{j_\ell}\}$. For $\mathbf{v} \in SR$ consisting of selectability sub-vector $\mathbf{s} \in S$ and rejectability sub-vector $\mathbf{r} \in \mathcal{R}$, we write $\mathbf{v} = (\mathbf{s}; \mathbf{r})$. Thus, $p_{SR}(\mathbf{v}) = p_{SR}(\mathbf{s}; \mathbf{r})$.

We extend this notation to conditional interdependence mass functions by defining $p_{SR\setminus S_1R_1|S_1R_1}(\mathbf{s}; \mathbf{r}|\mathbf{s}_1; \mathbf{r}_1)$ for $(\mathbf{s}; \mathbf{r}) \in \mathbf{U}_{SR\setminus S_1R_1}$ and $(\mathbf{s}_1; \mathbf{r}_1) \in \mathbf{U}_{S_1R_1}$ as the **conditional interdependence mass function** of $(\mathbf{s}; \mathbf{r})$ given that all interdependence mass over $\mathbf{U}_{S_1R_1}$ is ascribed to $(\mathbf{s}_1; \mathbf{r}_1)$. This conditional mass function permits the decision makers to deal with all possible hypothetical situations; that is, conditioned on each $(\mathbf{s}_1; \mathbf{r}_1) \in \mathbf{U}_{S_1R_1}$, the conditional mass function characterizes the joint selectability and rejectability of each $(\mathbf{s}; \mathbf{r}) \in \mathbf{U}_{SR\setminus S_1R_1}$. Letting $\mathbf{f} = (\mathbf{s}_1; \mathbf{r}_1) \in S_1R_1$ and $\mathbf{g} = (\mathbf{s}; \mathbf{r}) \in SR\setminus S_1R_1$, the transition relationship (6.3) becomes

$$P_{SR}(\mathbf{F} \times \mathbf{G}) = \sum_{\mathbf{f}\in\mathbf{F}} \sum_{\mathbf{g}\in\mathbf{G}} p_{SR\setminus S_1R_1|S_1R_1}(\mathbf{g}|\mathbf{f}) p_{S_1R_1}(\mathbf{f}). \qquad (6.5)$$

When $\mathbf{F} = \{\mathbf{f}\} = \{(\mathbf{s}_1; \mathbf{r}_1)\} \subset \mathcal{SR}$ and $\mathbf{G} = \{\mathbf{g}\} = \{(\mathbf{s}; \mathbf{r})\} \subset \mathcal{SR} \backslash \mathcal{S}_1 \mathcal{R}_1$ are singleton sets, we write (6.5) as

$$p_{\mathcal{SR}}(\mathbf{s}, \mathbf{s}_1; \mathbf{r}, \mathbf{r}_1) = p_{\mathcal{SR} \backslash \mathcal{S}_1 \mathcal{R}_1 | \mathcal{S}_1 \mathcal{R}_1}(\mathbf{s}; \mathbf{r} | \mathbf{s}_1; \mathbf{r}_1) \cdot p_{\mathcal{S}_1 \mathcal{R}_1}(\mathbf{s}_1; \mathbf{r}_1). \tag{6.6}$$

We may factor the interdependence function $p_{\mathcal{S}_1 \mathcal{R}_1}(\mathbf{s}_1; \mathbf{r}_1)$ further by defining the sub-intermixture $\mathcal{S}_2 \mathcal{R}_2$ of $\mathcal{S}_1 \mathcal{R}_1$, as follows. Let $(\mathbf{s}_1; \mathbf{r}_1) = (\mathbf{s}', \mathbf{s}_2; \mathbf{r}', \mathbf{r}_2)$, where $(\mathbf{s}'; \mathbf{r}') \in \mathbf{U}_{\mathcal{S}_1 \mathcal{R}_1 \mathcal{S}_2 \mathcal{R}_2}$ and $(\mathbf{s}_2; \mathbf{r}_2) \in \mathbf{U}_{\mathcal{S}_2 \mathcal{R}_2}$. Then $p_{\mathcal{S}_1 \mathcal{R}_1}(\mathbf{s}_1; \mathbf{r}_1) = p_{\mathcal{S}_1 \mathcal{R}_1 \backslash \mathcal{S}_2 \mathcal{R}_2 | \mathcal{S}_2 \mathcal{R}_2}(\mathbf{s}'; \mathbf{r}' | \mathbf{s}_2; \mathbf{r}_2) \cdot p_{\mathcal{S}_2 \mathcal{R}_2}(\mathbf{s}_2; \mathbf{r}_2)$. Substituting this into (6.6) yields

$$p_{\mathcal{SR}}(\mathbf{s}, \mathbf{s}', \mathbf{s}_2; \mathbf{r}, \mathbf{r}', \mathbf{r}_2) = p_{\mathcal{SR} \backslash \mathcal{S}_1 \mathcal{R}_1 | \mathcal{S}_1 \mathcal{R}_1}(\mathbf{s}; \mathbf{r} | \mathbf{s}', \mathbf{s}_2; \mathbf{r}', \mathbf{r}_2)$$
$$\cdot p_{\mathcal{S}_1 \mathcal{R}_1 \backslash \mathcal{S}_2 \mathcal{R}_2 | \mathcal{S}_2 \mathcal{R}_2}(\mathbf{s}'; \mathbf{r}' | \mathbf{s}_2; \mathbf{r}_2) \cdot p_{\mathcal{S}_2 \mathcal{R}_2}(\mathbf{s}_2; \mathbf{r}_2). \tag{6.7}$$

Equation (6.7) is known as the *chain rule of probability theory* (Eisen, 1969). This rule permits the global expression of the interdependence function as the product of local conditional relationships. Just as it is often easier to compose a joint probability distribution from conditional distributions (e.g., Markov processes), it is conceptually easier to construct joint interdependencies from conditional interdependencies than to construct the joint interdependencies directly.

Definition 6.9

Let $\mathbf{U} = U_1 \times \cdots \times U_N$, where each U_i is a Euclidean space consisting of a continuum of elements (e.g., $U_i \subset \mathbb{R}^n$), let $\mathcal{F}_i$ be the Borel field over U_i and let $P_{\mathcal{SR}}$ be the interdependence measure for the intermixture $\mathcal{SR}$. Furthermore, suppose $P_{\mathcal{SR}}$ is absolutely continuous with respect to Lebesgue measure over $\mathbf{U}_{\mathcal{SR}}$. The **interdependence density function** is a Lebesgue measurable function, $p_{\mathcal{SR}}$, that satisfies the property

$$P_{\mathcal{SR}}(\mathbf{V}) = \int_{\mathbf{V}} p_{\mathcal{SR}}(\mathbf{v}) d\mathbf{v}$$

for all measurable $\mathbf{V} \subset \mathbf{U}_{\mathcal{SR}}$; that is, $p_{\mathcal{SR}}$ is the Radon–Nikodym derivative of $P_{\mathcal{SR}}$ with respect to Lebesgue measure. The transition relationship (6.3) becomes

$$P_{\mathcal{SR}}(\mathbf{F} \times \mathbf{G}) = \int_{\mathbf{F}} \int_{\mathbf{G}} p_{\mathcal{SR} \backslash \mathcal{S}_1 \mathcal{R}_1 | \mathcal{S}_1 \mathcal{R}_1}(\mathbf{g} | \mathbf{f}) p_{\mathcal{S}_1 \mathcal{R}_1}(\mathbf{f}) \, d\mathbf{g} \, d\mathbf{f}.$$

Letting $\mathbf{f} = (\mathbf{s}_1; \mathbf{r}_1) \in \mathcal{S}_1 \mathcal{R}_1$ and $\mathbf{g} = (\mathbf{s}; \mathbf{r}) \in \mathcal{SR} \backslash \mathcal{S}_1 \mathcal{R}_1$, the interdependence density becomes

$$p_{\mathcal{SR}}(\mathbf{s}, \mathbf{s}_1; \mathbf{r}, \mathbf{r}_1) = p_{\mathcal{SR} \backslash \mathcal{S}_1 \mathcal{R}_1 | \mathcal{S}_1 \mathcal{R}_1}(\mathbf{s}; \mathbf{r} | \mathbf{s}_1; \mathbf{r}_1) \cdot p_{\mathcal{S}_1 \mathcal{R}_1}(\mathbf{s}_1; \mathbf{r}_1)$$

where $p_{\mathcal{SR} \backslash \mathcal{S}_1 \mathcal{R}_1 | \mathcal{S}_1 \mathcal{R}_1}(\mathbf{s}; \mathbf{r} | \mathbf{s}_1; \mathbf{r}_1)$ is the **conditional interdependence density function**.

$\square$

To illustrate, let $\mathbf{X} = \{X_1, X_2, X_3\}$ and let $\mathcal{S} = S_1 S_2$ and $\mathcal{R} = R_3$. Then $\mathcal{SR} = S_1 S_2 R_3$, $\mathbf{SR} \setminus \mathcal{SR} = S_3 R_1 R_2$. The interdependence measure is

$$P_{S_1 S_2 S_3 R_1 R_2 R_3}(V_1 \times V_2 \times V_3 \times W_1 \times W_2 \times W_3) =$$

$$\int_{V_1 \times V_2 \times W_3} P_{S_3 R_1 R_2 | S_1 S_2 R_3}(V_3 \times W_1 \times W_2 | v_1, v_2, w_3)$$

$$\cdot P_{S_1 S_2 R_3}(dv_1 \times dv_2 \times dw_3). \tag{6.8}$$

Now let $S_1 R_1 = S_2 R_3$, so $\mathcal{SR} \setminus S_1 R_1 = S_1$. Then

$$P_{S_1 S_2 S_3 R_1 R_2 R_3}(V_1 \times V_2 \times V_3 \times W_1 \times W_2 \times W_3) =$$

$$\int_{V_1 \times V_2 \times W_3} P_{S_3 R_1 R_2 | S_1 S_2 R_3}(V_3 \times W_1 \times W_2 | v_1, v_2; w_3)$$

$$\cdot P_{S_1 | S_2 R_3}(dv_1 | v_2; w_3) \cdot P_{S_2 R_3}(dv_2 \times dw_3). \tag{6.9}$$

If the option space is discrete, the interdependence mass function is

$$p_{S_1 S_2 S_3 R_1 R_2 R_3}(v_1, v_2, v_3; w_1, w_2, w_3) = p_{S_3 R_1 R_2 | S_1 S_2 R_3}(v_3; w_1, w_2 | v_1, v_2; w_3)$$

$$\cdot p_{S_1 | S_2 R_3}(v_1 | v_2; w_3) p_{S_2 R_3}(v_2; w_3). \tag{6.10}$$

The function $p_{S_3 R_1 R_2 | S_1 S_2 R_3}(v_3; w_1, w_2 | v_1, v_2; w_3)$ is the conditional interdependence associated with X_3 selecting option v_3, X_1 rejecting option w_1, and X_2 rejecting option w_2, conditioned on X_1 ascribing its entire unit of selectability mass on v_1, X_2 ascribing its entire unit of selectability mass on v_2, and X_3 ascribing its entire unit of rejectability mass on w_3. In other words, suppose X_1 views v_1 as totally avoiding failure, X_2 views v_2 as the total absence of conservation, and X_3 views w_3 also as the total absence of conservation; then $p_{S_3 R_1 R_2 | S_1 S_2 R_3}(v_3; w_1, w_2 | v_1, v_2; w_3)$ characterizes the conditional interdependence associated with X_3 selecting v_3, X_1 rejecting w_1, and X_2 rejecting w_2. When the option space is continuous and densities exist, the interdependence density function assumes the same form as (6.10), with all functions being interpreted as densities, rather than as mass functions.

Conditional interdependence permits the expression of situational altruism (see Section 1.3.2), whereby a decision maker may accommodate the interests of another decision maker by adjusting its selectability or rejectability *given that the other decision maker places its selectability or rejectability on certain options.*

Example 6.2 We continue with Example 6.2 with Lucy $= X_L$ and Ricky $= X_R$, and consider R $= R_L R_R$, the joint rejectability mixture. The corresponding joint rejectability function $p_{R_L R_L}(r_L, r_R)$ may be factored to become

$$p_{R_L R_R}(r_L, r_R) = p_{R_L | R_R}(r_L | r_R) p_{R_R}(r_R).$$

Each option $r_R \in U_R = \{T, F\}$ generates a different conditional rejectability mass function $p_{R_L | R_R}(\cdot | r_R)$; that is, $p_{R_L | R_R}(r | r_R) \geq 0$ for all $r \in U_L = \{P, H\}$ and $\sum_{r \in U_L} p_{R_L | R_R}(r | r_R) = 1$. The joint interaction space is then

$$U = U_L \times U_R = \{P, H\} \times \{T, F\}.$$

Now, suppose that Lucy not adopting a certain option, say P, would be beneficial to Ricky if he were to favor one of his options, say F, but if he were not to favor F, then Lucy would not be influenced by Ricky's interests. Lucy could accommodate this situation by the following conditional rejectability structure.

$$p_{R_L|R_R}(r_L|r_R) = \begin{cases} \begin{cases} 1 & \text{if } r_L = P \text{ and } r_R \neq F, \\ 0 & \text{if } r_L \neq P \text{ and } r_R \neq F, \end{cases} \\ p'_{R_L}(r_L) \text{ if } r_R = F, \end{cases} \tag{6.11}$$

where p'_{R_L} is Lucy's rejectability based solely on her own interests without taking into consideration Ricky's desires. Notice that Lucy can specify this conditional rejectability without knowing Ricky's actual rejectability structure – it is a purely hypothetical consideration. If Ricky were indeed to place low rejectability on F, and hence high rejectability on the complement of F, by setting $p_{R_R}(T) \approx 1$, then Lucy would accommodate that situation and her individual rejectability of P would become

$$p_{R_L}(P) = \sum_{s \in U_L} p_{R_L|R_R}(P|R_R) p_{R_L}(s) \approx \begin{cases} 1 & \text{if } p_{R_L}(T) \approx 1, \\ p'_{R_L}(P) & \text{if } p_{R_R}(T) \ll 1. \end{cases} \tag{6.12}$$

In this way, Lucy exhibits situational altruism; that is, she is willing to accommodate Ricky by rejecting P if, but only if, doing so is critical to Ricky's welfare. If, however, at the moment of commitment, Ricky does not in fact have a strong preference for F, the opportunity to take advantage of Lucy's largesse is not exercised and Lucy is not vulnerable to needless extra cost.

Clearly, conditional preferences are neutral. Instead of situational benevolence, a decision maker may practice situational malevolence by adjusting its conditional preferences in ways designed to injure others, even at its own expense. In fact, a decision maker can both reward and punish another's behavior by appropriately defining its conditional preferences. Once we eschew the strict doctrine of individual rationality, there is no built-in bias associated with preference relationships between decision makers.

6.2.3 Spatial emergence

Although each of the conditional mass functions used to create the interdependence function represents a total ordering, it is a *local* total ordering and involves only a subset of agents and concerns. *Each of these local total orderings is only a partial ordering, however, if viewed from the global, or community-wide, perspective. By combining such local total orderings together according to the chain rule of probability, a global total ordering emerges from the local orderings. The joint selectability and rejectability mass functions then characterize emergent global behavior and the individual selectability and rejectability marginals characterize emergent individual behavior.* Thus, both individual and group behavior emerge as consequences of local conditional interests that propagate throughout the community from the interdependent local to the interdependent global and from the conditional to the unconditional.

As noted by Pearl (1988), conditional probabilities permit local, or specific, responses to be characterized; they possess modularity features similar to logical production rules. Conditional behavior is behavior at the local level, that is, with all relevant circumstances specified. By factoring the interdependence function into products of conditional interdependencies, we are able to characterize the global relationships in terms of local relationships, which are often easier to specify. In effect, the interdependence function provides a mechanism for implementing some types of production rules.

The interdependence function possesses several properties that make it a useful and efficient way of characterizing relationships that may exist between decision makers: First, it provides an evaluation of every possible consideration that could be relevant to the decision, when viewed jointly from the perspectives of avoidance of failure and conservation of resources. Second, it contains no redundant information. In this sense, it is a parsimonious characterization of all possible failure avoidance/resource conservation situations that can occur. Finally, it guarantees that local, or conditional, assessments of failure avoidance and resource conservation are consistent with global, or unconditional, combined assessments. Contradictory assessments cannot exist.

Let us refer to the interdependence measure, the interdependence mass function, and the interdependence density function all as interdependence functions, and let context determine which of these quantities is relevant.

Example 6.3 We are now in a position to specify the interdependence function for Lucy's and Ricky's car-buying adventure in Example 6.1. This function is of the form $p_{S_L S_R R_L R_R}(u_L, u_R; v_L, v_R)$, where

$$(u_L, u_R; v_L, v_R) \in U \times U = \{\{P, H\} \times \{T, F\}\} \times \{\{P, H\} \times \{T, F\}\}.$$

Let us begin by expressing the interdependence function as the product of conditional selectability and rejectability functions.

$$p_{S_L S_R R_L R_R}(u_L, u_R; v_L, v_R) = p_{S_L|S_R R_L R_R}(u_L|u_R; v_L, v_R) \cdot p_{S_R|R_L R_R}(u_R|v_L, v_R)$$
$$\cdot p_{R_L|R_R}(v_L|v_R) \cdot p_{R_R}(v_R). \tag{6.13}$$

We may invoke some obvious simplifications at this point by noting that any single decision maker's selectability will not depend on its rejectability, so we may rewrite (6.13) as

$$p_{S_L S_R R_L R_R}(u_L, u_R; v_L, v_R) = p_{S_L|S_R R_R}(u_L|u_R, v_R) \cdot p_{S_R|R_L}(u_R|v_L)$$
$$\cdot p_{R_L|R_R}(v_L|v_R) \cdot p_{R_R}(v_R). \tag{6.14}$$

p_{R_R} is Ricky's myopic rejectability and $p_{R_L|R_R}$ is Lucy's rejectability conditioned on Ricky's rejections. Also, $p_{S_R|R_L}$ is Ricky's selectability conditioned on Lucy's rejections and $p_{S_L|S_R R_R}$ is Lucy's selectability conditioned on Ricky's selections and rejections. For this simple two-choice problem, rejecting one option means selecting the other, so we may simplify $p_{S_L|S_R R_R}$ to become $p_{S_L|R_R}$, yielding a factorization of the interdependence function of the form

$$p_{S_L S_R R_L R_R}(u_L, u_R; v_L, v_R) = p_{S_L|R_R}(u_L|v_R) \cdot p_{S_R|R_L}(u_R|v_L) \cdot p_{R_L|R_R}(v_L|v_R) \cdot p_{R_R}(v_R). \tag{6.15}$$

Let us begin with the specification of Ricky's myopic rejectability. Since there are only two choices, his rejectability assignment must be of the form

$$p_{R_R}(T) = \beta,$$
$$p_{R_R}(F) = 1 - \beta,$$

where, in the interest of conserving resource (room for passengers), we may assume that $0 < \beta < 0.5$.

We next consider $p_{S_L|R_R}$. Given that Ricky rejects T, Lucy's obvious response would be to place all of her selectability mass on H (since Porches are not available in four-door models). Given that Ricky rejects F, she would place α worth of selectability on P and $1 - \alpha$ on H. Let us assume that Lucy considers a Porsche to be more stylish than a Honda, which means that we restrict $1 \geq \alpha > 0.5$. Lucy's conditional selectability function $p_{S_L|R_R}$ would thus be of the form

	v_R	
u_L	T	F
P	0	α
H	1	$1-\alpha$

To specify $p_{S_R|R_L}$, note that if Lucy rejects H, then Ricky should place all of his selectability mass on T. If, on the other hand, Lucy rejects P, then Ricky's myopic preferences are operative and his conditional selectability function, $p_{S_R|R_L}$, would be

	v_L	
u_R	P	H
T	β	1
F	$1-\beta$	0

Finally, to specify $p_{R_L|R_R}$, let us suppose that Lucy is willing to offer some deference to Ricky's interests and is willing to adopt a stance of situational altruism. Accordingly, she incorporates the conditional rejectability defined by (6.11), with

$$p'_{R_L}(P) = 1 - \alpha,$$
$$p'_{R_L}(H) = \alpha.$$

The resulting conditional rejectability function is

	v_R	
v_L	T	F
P	1	$1-\alpha$
H	0	α

Substituting these functions in to (6.15) yields the interdependence function displayed in Table 6.2. The next step is for the participants to obtain a multiple-agent satisficing solution based upon these values.

6.3   Satisficing games

Praxeic utility theory was developed for an abstract option space of arbitrary dimensionality; all that is required is that the selectability and rejectability measures be defined. Consequently, the theory immediately extends to the multi-agent case by considering multipartite option spaces. The role of the interdependence function is to characterize all of the relevant relationships that exist among the members of the multi-agent

Table 6.2: The interdependence function for Lucy and Ricky

(u_L, u_R, v_L, v_R)	$p_{S_L S_R R_L R_R}$	(u_L, u_R, v_L, v_R)	$p_{S_L S_R R_L R_R}$
(P, T, P, T)	0	(H, T, P, T)	β^2
(P, T, P, F)	$\alpha\beta(1 - \alpha)(1 - \beta)$	(H, T, P, F)	$(1 - \alpha)^2\beta(1 - \beta)$
(P, T, H, T)	0	(H, T, H, T)	0
(P, T, H, F)	$\alpha^2(1 - \beta)$	(H, T, H, F)	$\alpha(1 - \alpha)(1 - \beta)$
(P, F, P, T)	0	(H, F, P, T)	$\beta(1 - \beta)$
(P, F, P, F)	$\alpha(1 - \alpha)(1 - \beta)^2$	(H, F, P, F)	$(1 - \alpha)^2(1 - \beta)^2$
(P, F, H, T)	0	(H, F, H, T)	0
(P, F, H, F)	0	(H, F, H, F)	0

community. Once this function is defined, the joint selectability and rejectability functions are obtainable via (6.1) and (6.2), resulting in

$$p_S(\mathbf{u}) = \sum_{\mathbf{v} \in \mathbf{U}} p_{SR}(\mathbf{u}; \mathbf{v}), \tag{6.16}$$

$$p_R(\mathbf{v}) = \sum_{\mathbf{u} \in \mathbf{U}} p_{SR}(\mathbf{u}; \mathbf{v}) \tag{6.17}$$

for the discrete case, and

$$p_S(\mathbf{u}) = \int_{\mathbf{U}} p_{SR}(\mathbf{u}; \mathbf{v}) d\mathbf{v}, \tag{6.18}$$

$$p_R(\mathbf{v}) = \int_{\mathbf{U}} p_{SR}(\mathbf{u}; \mathbf{v}) d\mathbf{u} \tag{6.19}$$

for the continuous case.

Definition 6.10

Let $\{X_1, \ldots, X_N\}$ be a set of N decision makers, let U_i be the option space corresponding to X_i, and let $\mathbf{U} = U_1 \times \cdots \times U_N$ be the joint option space. A **satisficing game** is the triple $\{\mathbf{U}, p_S, p_R\}$, where p_S is the joint selectability function and p_R is the joint rejectability function.

The **joint solution** to the satisficing game $\{\mathbf{U}, p_S, p_R\}$ with boldness q is the set

$$\Sigma_q = \{\mathbf{u} \in \mathbf{U}: p_S(\mathbf{u}) \geq q p_R(\mathbf{u})\}. \tag{6.20}$$

Σ_q is termed the **jointly satisficing set**, and elements of Σ_q are **jointly satisficing options**. Equation (6.20) is the **joint praxeic likelihood ratio test** (JPLRT). □

The set of individually satisficing options for each player is obtained by computing the marginal selectability and rejectability functions for each X_i, $i = 1, \ldots, N$ from

(6.16) and (6.17), yielding

$$p_{S_i}(u_i) = \sum_{\substack{u_j \in U_j \\ j \neq i}} p_{S_1 \cdots S_N}(u_1, \dots, u_N), \tag{6.21}$$

$$p_{R_i}(u_i) = \sum_{\substack{u_j \in U_j \\ j \neq i}} p_{R_1 \cdots R_N}(u_1, \dots, u_N) \tag{6.22}$$

for $i = 1, \dots, N$.

Definition 6.11

The **individual solutions** to the satisficing game $\{\mathbf{U}, p_{\mathbf{S}}, p_{\mathbf{R}}\}$ are the sets

$$\Sigma_q^i = \{u_i \in U_i \colon p_{S_i}(u_i) \geq q p_{R_i}(u_i)\}, \tag{6.23}$$

$i = 1, \dots, N$. Σ_q^i is the **individually satisficing set** for X_i, $i = 1, \dots, N$. The product set consisting of the individually satisficing sets is the **satisficing rectangle**:

$$\mathfrak{R}_q = \Sigma_q^1 \times \cdots \times \Sigma_q^N = \{(u_1, \dots, u_N) \colon u_i \in \Sigma_q^i, i = 1, \dots, N\}. \tag{6.24}$$

$\square$

Example 6.4 We may now compute the joint and individually satisficing decisions for Lucy and Ricky in Example 6.1. In the interest of simplicity, let us impose the constraint that $\beta = 1 - \alpha$, that is, Ricky's myopic rejectability of two seats is equal to Lucy's non-altruistic rejectability of Porsche. This restriction will reduce the complexity of the following expressions without reducing the pedagogical value of the example. To compute the jointly satisficing set, we apply (6.16) and (6.17) to obtain

$$P_{S_L S_R}(P, T) = \alpha^2(1 - \alpha)^2 + \alpha^3,$$
$$P_{S_L S_R}(P, F) = \alpha^3(1 - \alpha),$$
$$P_{S_L S_R}(H, T) = (1 - \alpha)^2 + \alpha(1 - \alpha)^3 + \alpha^2(1 - \alpha),$$
$$P_{S_L S_R}(H, F) = \alpha^2(1 - \alpha)^2 + \alpha(1 - \alpha)$$

and

$$P_{R_L R_R}(P, T) = 1 - \alpha,$$
$$P_{R_L R_R}(P, F) = \alpha(1 - \alpha),$$
$$P_{R_L R_R}(H, T) = 0,$$
$$P_{R_L R_R}(H, F) = \alpha^2.$$

From these values it is clear that (P, F) is never satisficing, (H, T) is always satisficing, and the satisficing status of (P, T) and (H, F) depend on the values of α. The jointly satisficing set is easily computed to be, for $q = 1$,

$$\Sigma_q = \begin{cases} \{\{H,T\},\{H,F\}\} & \text{for } 0.5 \leq \alpha \leq 0.555, \\ \{H,T\} & \text{for } 0.555 < \alpha < 0.660, \\ \{\{H,T\},\{P,T\}\} & \text{for } 0.660 \leq \alpha \leq 1. \end{cases} \tag{6.25}$$

We may compute the marginals for Lucy and Ricky as follows. For Lucy, we obtain

$$p_{S_L}(P) = \alpha^2,$$
$$p_{S_L}(H) = 1 - \alpha^2,$$

$$p_{R_L}(P) = 1 - \alpha^2,$$
$$p_{R_L}(H) = \alpha^2,$$

and for Ricky we obtain

$$p_{S_R}(T) = 1 - \alpha + \alpha^3,$$
$$p_{S_R}(F) = \alpha - \alpha^3,$$

$$p_{R_R}(T) = 1 - \alpha,$$
$$p_{R_R}(F) = \alpha.$$

Comparing these values with $q = 1$, we see that Lucy's individual satisficing set is

$$\Sigma_q^L = \begin{cases} \{P\} & \text{for } 0.707 \leq \alpha \leq 1, \\ \{H\} & \text{for } 0.5 \leq \alpha < 0.707, \end{cases}$$

and Ricky's individually satisficing set is

$$\Sigma_q^R = \{T\}.$$

Concatenating the individual interests of Lucy and Ricky, we obtain the satisficing rectangle

$$\mathfrak{R}_q = \Sigma_q^L \times \Sigma_q^R = \begin{cases} \{P,T\} & \text{for } 0.707 \leq \alpha \leq 1, \\ \{H,T\} & \text{for } 0.5 \leq \alpha < 0.707. \end{cases} \tag{6.26}$$

We see that when $0.5 \leq \alpha \leq 0.707$ the join option $\{H,T\}$ is both jointly and individually satisficing, and when $0.707 \leq \alpha \leq 1$, the option vector $\{P,T\}$ is both jointly and individually satisficing. The problem is parameterized by Ricky's myopic preferences regarding the number of passengers that can be accommodated, and by Lucy's style preferences.

6.4   Group preference

One of the issues that has perplexed game theorists is how to characterize group preferences. The root of the problem is that optimization at the group level cannot be made to be consistent with optimization at the individual level. But, if we replace the demand for optimization with an attitude of satisficing, both group and individual interests may emerge from a more holistic model of inter-agent relationships.

As we saw in Section 6.2.3, it is possible to combine the conditional interdependencies that exist between decision makers to form an interdependence function that accounts for all of the preference relationships that exist between members of a group. The fact that the interdependence function is able to account for conditional preference dependencies between decision makers provides a coupling between decision makers that permits them to widen their spheres of interest beyond their own myopic preferences. This widening of preferences leads to a concept of group preference.

Definition 6.12
The **satisficing group preference** at boldness q of a multi-agent system is the set of all option vectors such that joint selectability equals or exceeds the index of boldness times the joint rejectability; that is, it is the jointly satisficing set Σ_q.   □

This is a weak notion of preference. It does not imply that there is some coherent notion of "group good," although such an implication is certainly not ruled out. To interpret this notion of group preference further requires operational definitions of joint selectability and joint rejectability. For problems where these notions are explicitly defined, it is straightforward to say what it means to be good enough for the group. However, if the interdependence function comprises conditional selectabilities and rejectabilities, a coherent operational definition of group selectability and group rejectability, and hence group preference, may be difficult to ascertain from the product of these conditional preferences.

This observation may help to explain why the notion of group preference is so elusive. It would seem that the notion of "group preference" should convey the idea of harmonious behavior, such as the individuals pursuing some common goal. But, since the group preference is not an explicit aggregation of the individual preferences of the participants (although it may be implicit in the conditional preferences), nor is it imposed by a superplayer, it need not correspond to harmonious behavior.

The interdependence function comprises the totality of preference relationships that exist between the players of the game. These preferences may be conditional or unconditional, they may be cooperative or competitive, and they may be egoistic or altruistic. They may result in highly efficient goal-directed behavior or they may result in dysfunctional behavior. Thus, the notion of satisficing group preference is completely conflict/coordination neutral. With a competitive game the group preference may be to oppose one another, while for a cooperative game the group preference may be to coordinate.

There is no requirement or expectation that group preferences will be derived by aggregating individual preferences (a bottom-up approach) or via a superplayer to dictate choices to the individuals so as to insure that the group's goals are met (a top-down approach). If such structures naturally occur through the specification of the preferences (conditional or otherwise), they can be accommodated in the satisficing context.

Consider the example of Lucy and Ricky buying a car as discussed in Example 6.4. The various conditional preferences that appear in (6.15) propagate through the system to define a set of group preferences given by (6.25). Notice, however, that these group preferences are not derived as aggregations of individual preferences (in fact, Lucy's myopic preferences are never specified – only her conditional preferences are given), nor are they dictated by a superplayer.

If neither top-down nor bottom-up structures exist with the game, then coherent group preferences, if they exist, may be **emergent**, in the sense that they are determined by the totality of the linked preferences, and display themselves only as the links are forged. It is analogous to making a cake. The various ingredients (flour, sugar, water, heat, etc.) influence each other in complex ways, but it is not until they are all combined in proper proportions that a harmonious group notion of "cakeness" emerges. However, if the ingredients are not compatible, then no harmonious notion of cakeness can emerge.

The options that lie in the individually satisficing sets Σ_q^i define the preferences for the individuals. These preferences may be myopic, that is, specified unconditionally, as is required with conventional game theory, or they may emerge from conditional preferences as the marginal selectability and rejectability functions are derived from the joint selectability and rejectability functions.

As was seen in Example 6.4 involving the purchase of Lucy and Ricky's car, it is not generally true that the individual satisficing decisions and the joint decisions must be consistent in some sense, that is, that at least one of the conditions $\Sigma_q = \Re_q, \Sigma_q \subseteq \Re_q$, or $\Sigma_q \supseteq \Re_q$ will hold. Unfortunately, this is not the case. In fact, the two sets may be disjoint. The marginal selectability and rejectability functions will not, in general, yield the same decisions as will the joint selectability and rejectability functions, even if both $p_{S_1 \ldots, S_N}$ and $p_{R_1 \ldots, R_N}$ factor into products of the form $p_{S_1 \ldots, S_N} = p_{S_1} \cdots p_{S_N}$ and $p_{R_1 \ldots, R_N} = p_{R_1} \cdots p_{R_N}$. I will discuss this issue in more detail and describe ways to reconcile group and individual preferences in Section 7.2.

To illustrate further the emergence of individual and group preferences, let us now address the Pot-Luck Dinner problem that was introduced in Example 2.1. To examine this problem from the satisficing point of view, we first need to specify operational definitions for selectability rejectability. Although there is not a unique way to frame this problem, let us take rejectability as cost of the meal and take selectability as enjoyment of the meal. The interdependence function is a function of six independent variables and may be factored, according to the chain rule, as

$$p_{S_L S_C S_M R_L R_C R_M}(x, y, z; u, v, w) = p_{S_C|S_L S_M R_L R_C R_M}(y|x, z; u, v, w)$$
$$\cdot p_{S_L S_M R_L R_C R_M}(x, z; u, v, w), \tag{6.27}$$

where the subscripts L, C, and M correspond to Larry, Curly, and Moe, respectively. The term $p_{S_C|S_L S_M R_L R_C R_M}(y|x, z; u, v, w)$ expresses the selectability that Curly places on option y, given that Larry selects x and rejects u, that Curly rejects v, and that Moe selects z and rejects w. From the statement of the problem, we realize that, based on Larry's selectability, Curly's selectability is independent of all other considerations. Thus we can simplify the conditional selectability to obtain

$$p_{S_C | S_L S_M R_L R_C R_M}(y | x, z; u, v, w) = p_{S_C | S_L}(y | x).$$

Next, we apply the chain rule to the second term on the right-hand side of (6.27), which yields

$$p_{S_L S_M R_L R_C R_M}(x, z; u, v, w) = p_{S_L S_M | R_L R_C R_M}(x, z | u, v, w) \cdot p_{R_L R_C R_M}(u, v, w).$$

But, on the basis of Curly's rejectability, the joint selectability for Larry and Moe is independent of all other considerations, so

$$p_{S_L S_M | R_L R_C R_M}(x, z | u, v, w) = p_{S_L S_M | R_C}(x, z | v).$$

By making the appropriate substitutions, (6.27) becomes

$$p_{S_L S_C S_M R_L R_C R_M}(x, y, z; u, v, w) = p_{S_C | S_L}(y | x) \cdot p_{S_L S_M | R_C}(x, z | v) \cdot p_{R_L R_C R_M}(u, v, w). \tag{6.28}$$

We desire to obtain Σ_q, the jointly satisficing options for the group and Σ_q^L, Σ_q^C, and Σ_q^M, the individually satisficing option sets for Larry, Curly, and Moe, respectively. To do so, we must specify each of the components of (6.28). To compute $p_{S_C | S_L}$, recall that Curly prefers beef to chicken to pork by respective factors of 2 conditioned on Larry preferring soup, but that Curly is indifferent conditioned on Larry preferring salad. We may express these relationships by the conditional selectability functions:

$$p_{S_C | S_L}(beef | soup) = 4/7, \qquad p_{S_C | S_L}(beef | sald) = 1/3,$$
$$p_{S_C | S_L}(chkn | soup) = 2/7, \qquad p_{S_C | S_L}(chkn | sald) = 1/3,$$
$$p_{S_C | S_L}(pork | soup) = 1/7, \qquad p_{S_C | S_L}(pork | sald) = 1/3.$$

To compute $p_{S_L S_M | R_C}$, we recall, given that Curly views pork as completely rejectable, Moe views lemon custard pie as highly selectable and Larry is indifferent. Given that Curly views beef as completely rejectable, Larry views soup as selectable, and Moe is indifferent. Furthermore, given that Curly views chicken as completely rejectable, both Larry and Moe are indifferent. These relationships may be expressed as

$$p_{S_L S_M | R_C}(soup, lcst | pork) = 0.5,$$
$$p_{S_L S_M | R_C}(soup, bcrm | pork) = 0.0,$$
$$p_{S_L S_M | R_C}(sald, lcst | pork) = 0.5,$$
$$p_{S_L S_M | R_C}(sald, bcrm | pork) = 0.0,$$

$$p_{S_L S_M | R_C}(soup, lcst | beef) = 0.5,$$
$$p_{S_L S_M | R_C}(soup, bcrm | beef) = 0.5,$$
$$p_{S_L S_M | R_C}(sald, lcst | beef) = 0.0,$$
$$p_{S_L S_M | R_C}(sald, bcrm | beef) = 0.0,$$

and

$$p_{S_L S_M | R_C}(soup, lcst|chkn) = 0.25,$$
$$p_{S_L S_M | R_C}(soup, bcrm|chkn) = 0.25,$$
$$p_{S_L S_M | R_C}(sald, lcst|chkn) = 0.25,$$
$$p_{S_L S_M | R_C}(sald, bcrm|chkn) = 0.25.$$

Lastly, we need to specify $p_{R_L R_C R_M}$, the joint rejectability function. This is done by normalizing the meal cost values in Table 2.1 by the total cost of all meals (e.g., $p_{R_L R_C R_M}(soup, beef, lcst) = 23/296$).

With the interdependence function so defined and letting $q = 1$, the jointly and individually satisficing meals are as displayed in Table 6.3 along with the attitude specific to each choice. Each of these option vectors is good enough for the group, considered as a whole. Figure 6.1 provides a cross-plot of joint rejectability and selectability; the satisficing combinations are the ones that lie above the line $q = 1$, as labeled. The horizontal and vertical lines represent the uniform distribution lines ($\frac{1}{n} = \frac{1}{12}$). We can easily see that the meal $\{soup, beef, bcrm\}$ is not an equilibrium choice, since it is more rejectable and less selectable than $\{soup, beef, lcst\}$.

The individually satisficing items, as obtained by computing the selectability and rejectability marginals, are also provided in Table 6.3 to be *soup, beef,* and *lemon custard.* Notably, this set of choices is also jointly satisficing. Thus, all of the individual preferences are intact at a reasonable cost and, if pie throwing should ensue, we may assume that it would be for recreation only and not retribution.

The joint rejectability is almost uniform, with $H(p_{R_L R_C R_M}) = 3.57$ bits (the entropy of the uniform distribution is 3.58 bits). Joint selectability, although somewhat less uniform, is still high, with $H(p_{S_L S_C S_M}) = 3.27$ bits. These numbers mean that this system has the ability to produce a meal that is adequate but not greatly pleasing and that there are no bargains. This sense is reinforced by Table 6.3 and Figure 6.1, where we see that the first two jointly satisficing options listed are both gratifying (but just barely), the third is ambivalent, and the fourth is dubious. Individually, Larry and Moe are gratified, but Curly is ambivalent.

With the Pot-Luck Dinner, we see that, although total orderings for neither individuals nor the group are specified, we can use the *a priori* partial preference orderings from the problem statement to generate emergent, or *a posteriori*, group and individual orderings. This is the inside-out, or meso-to-micro/macro view of spatial emergence that was discussed in Section 2.1. For this example, a harmonious interpretation of group preference, namely, the avoidance of conflict, can be associated with the jointly satisficing set, since the two jointly gratifying meals fully conform to the preferences included in the problem statement. However, this group desideratum was *not* specified *a priori,* nor was harmony between group and individual preferences guaranteed. The players of this game are fortunate that the intersection of the jointly satisficing set Σ_q and the satisficing rectangle $\mathfrak{R}_q$ is not empty. If it were, then the group would be

Table 6.3: Jointly and individually satisficing choices for the Pot-Luck Dinner

Jointly satisficing			
Meal	$p_{S_L S_C S_M}$	$p_{R_L R_C R_M}$	Attitude
{soup, beef, lcst}	0.237	0.078	Gratification
{soup, chkn, lcst}	0.119	0.074	Gratification
{soup, beef, bcrm}	0.149	0.091	Ambivalence
{sald, pork, lcst}	0.080	0.074	Dubiety

Individually satisficing				
Participant	Choice	p_S	p_R	Attitude
Larry	soup	0.676	0.480	Gratification
Curly	beef	0.494	0.351	Ambivalence
Moe	lcst	0.655	0.459	Gratification

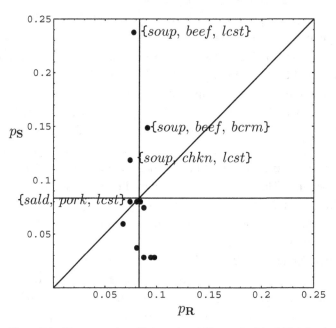

Figure 6.1: The cross-plot of joint rejectability and selectability for the Pot-Luck Dinner.

dysfunctional. In Section 7.2 I describe a way for the players of a satisficing game to negotiate when jointly satisficing and individually satisficing sets are incompatible.

A posteriori individual orderings also emerged from this exercise: Larry prefers soup to salad, Moe prefers lemon custard pie to banana cream pie, and Curly prefers beef to either chicken or pork. Note, however, that Curly is not required to impose a total ordering on his preferences (chicken versus pork). This approach does not force the generation of unwarranted preference relationships.

These attitudes (as defined in Section 5.4) are also apparent from calculations of diversity. The diversity values are 0.113, 0.070, and 0.112 for Larry, Curly, and Moe, respectively. Curly's selectability and rejectability are more nearly aligned than are either Larry's or Moe's, indicating that this decision problem is somewhat more difficult, or stressful, for Curly than for either of his partners. Relative to his dinner mates, Curly is ambivalent. This situation might prompt Curly to be in better touch with his own preferences or to find out more about the preferences of his partners before deciding, or both.

6.5　Optimizing versus satisficing

Von Neumann–Morgenstern utility functions are designed to express the preferences of individuals as a function only of the possible options chosen by themselves and others. This restriction in design is necessary if the outcome is to be consistent with the principle of individual self-interest. It is only because other players can influence a given player's payoff that a player is obligated, in defense of its own self-interest, to consider the interests of others as well as its own. Thus, although a self-interested player must take into account other player's utilities when choosing its strategy, it has no need or desire to consider the interests of other players when defining its own interests. This stance fosters competitive behavior, regardless of whether the game is conflictive or cooperative in nature.

Satisficing game theory, on the other hand, is completely neutral with regard to conflictive and coordinative aspects of the game; both aspects can be accommodated by appropriately structuring the interdependence function. Selectability and rejectability do not favor either aspect. With competitive games, conflict can be introduced through conditional selectability and rejectability functions that account for the differences in goals and values of the players. With cooperative games, coordinated behavior can be introduced through the same procedure.

The use of interdependence functions in place of von Neumann–Morgenstern utilities is an important difference between von Neumann–Morgenstern games and satisficing games. The interdependence function does not measure the amount of utility that accrues to a player as a result of a joint option being taken. Rather, it measures the preferences of all players as a function of their preferences for selecting and rejecting various options. If desired, the strength of the preferences can be made proportional to the payoff values.

A distinctive feature of satisficing theory as I have developed it is the incorporation of two distinct utility functions. These normalized utilities, denoted selectability and rejectability, are used to compare the positive and negative aspects of each option. The usage of such "dual utilities" has also been advocated by Margolis (1990), who advocates the use of two distinct utilities to characterize group and individual interests.[4] Margolis presents the view that a decision maker in a group context will possess a *social utility* and a *private utility*, and will adjust the assignment of its resources to achieve a balance between the two utility functions. This approach is essentially a special case of satisficing. Assuming that explicit utilities for both the group and the individuals were available (which assumption is not required for the satisficing approach), we could frame a decision problem such that, say, selectability represented the private utilities of selecting individual options and rejectability represented the social utility of rejecting a group option. Margolis' model would then comply with the satisficing approach.

Optimization is a strongly entrenched procedure and dominates conventional decision-making methodologies. There is great comfort in following traditional paths, especially when those paths are founded on such a rich and enduring tradition as rational choice affords. But, when synthesizing an artificial system, the designer has the opportunity to impose upon the agents a more socially accommodating paradigm. In particular, the designer of an artificial system must accommodate all of the relevant inter-relationships that exist between the participants. The set of possible interactions between decision makers is rich and diverse, as indicated by the following list of the major forms of group organization. The reader may judge whether or not individual rationality is an adequate model of behavior for these forms.

Definition 6.13

X_i **conflicts** with X_j if X_i's avoidance of failure or resource conservation interests are in opposition to those of X_j.

X_i **exploits** X_j if X_i is able to influence X_j to increase its avoidance of failure or decrease its conservation of resources without regard for those of X_j.

X_j **submits** to X_i if X_j acts in a way to increase the avoidance of failure or increase the conservation of resources for X_i without regard for its own failure avoidance or resource conservation.

X_i is **indifferent** to X_j if X_j's existence is completely irrelevant to X_i (though not necessarily vice versa).

X_i **tolerates** X_j if X_i acknowledges the existence of X_j but X_i's failure avoidance and resource conservation are independent from those of X_j, and the decision makers neither reinforce nor inhibit each other in any systematic way.

X_i **accommodates** X_j if X_i seeks to increase failure avoidance or increase resource conservation for X_j so long as it does not decrease failure avoidance or decrease resource conservation for itself.

---

[4] Other researchers have also advocated the use of multiple utilities. For example, Harsanyi's "ethical" and "subjective" preferences distinguish between purely social considerations and personal interests (Harsanyi, 1955; Sen, 1990), but Margolis' development is, to my knowledge, the only other attempt to incorporate such utilities in a way that is not based upon the premises of classical rational choice.

X_i **coordinates** with X_j if X_i's failure avoidance and resource conservation interests are compatible with those of X_j.   □

Any mixture of these preference relationships may exist in any given game and all such preferences may be expressed via the interdependence function. Note that this list does not include the term "cooperation." Cooperation has a special connotation in von Neumann–Morgenstern game theory. Decision makers are said to cooperate if they enter into binding coalitions; that is, they agree before decisions are made that, at the moment of truth, they will function as a unit. Much of N-person game theory deals with the ways in which coalitions may be justified. Chapter 7 discusses in detail the role of coalitions in multi-agent decision making and argues that coalitions should naturally emerge, if they arise at all, from the structure of the inter-agent preference relationships, rather than being imposed by logic that makes up for the shortcomings of the basic premises used to justify the choices.

7 Congruency

The enormous potential in mutual benefit (cooperative) strategies will not be tapped – or even understood – until we broaden our perspective beyond the narrow prejudice that we always do best by trying to beat others.

Alfie Kohn
No Contest (Houghton Mifflin, 1986)

Unless a community is a total dictatorship or is given to complete anarchy, there will be some form of sociality that is conducive to at least a weak form of agreement-seeking, or congruency. Cooperative societies may be expected to agree to work together, competitive societies may be expected to agree to oppose each other, and mixed-motive societies may be expected to agree to compromises that balance their interests. The procedures used to arrive at these agreements, however, are not determined simply as a function of the preference structure of the decision makers, either for von Neumann–Morgenstern scenarios or for satisficing scenarios. Instead, the procedures often involve some form of negotiation. Negotiation is a deliberative process whereby multiple decision makers can evaluate and share information when they have incentives to strike a mutually acceptable compromise.

In this chapter we first review von Neumann–Morgenstern N-person game theory, which forms the basis of classical negotiation theory. We then describe a new approach to negotiation based on satisficing game theory.

7.1 Classical negotiation

With the von Neumann–Morgenstern approach to decision making, the decision makers must each choose a strategy that represents an equilibrium such that it cannot improve its expected satisfaction by unilaterally altering its choice. Thus, the problem is reduced to one of identifying the equilibrium point (or points – there may be more than one). To define the equilibria it is necessary to abstract the game from its context and to capture all relevant information in the utility functions. By so doing, the game becomes an array of expected utilities and the context is reduced merely to a story line that accompanies the game to give it appeal and that also allows the solution, once it is obtained, to be interpreted. The story line itself has nothing to do with generating the solution. Once the equilibria are identified, the game is solved. The procedure used to solve it remains in the background. Optimization, by its nature, is concerned only with outcomes – it is insensitive to the process of obtaining a solution.

This insensitivity presents a problem, however, in negotiation due to the difficulty of dealing with the dynamic nature of coalition formation. Consequently, much of classical game theory has been focused on situations where the process of negotiation can be presumed irrelevant to actual play. In other words, all of the deals, promises, threats, etc., are presumed to take place before the first move is actually played (Shubik, 1982).

One of the great challenges to von Neumann–Morgenstern game theory is to balance individual preferences in a way that is consistent with any group preferences that are conceivable without an explicit notion of group preference. An exclusive reliance on individual rationality greatly limits the opportunity to account for group interest. Seeking a Pareto equilibrium is often suggested as a way to account for the interests of the group. The success of this application presupposes, however, that the players are disposed to be cooperative. But no such supposition is *a priori* justified by individual rationality. Any such dispositions are extra-game-theoretic, that is, if they do not influence the utility functions, they are not part of the mathematical game, no matter what story line accompanies the payoff array. The Prisoner's Dilemma game (see Section 8.1.3) is an excellent example. Although mutual cooperation is a Pareto equilibrium, under the von Neumann–Morgenstern version of the game, the players have absolutely no incentive to do so – indeed, they have a strong incentive not to cooperate. Cooperative behavior cannot be justified or motivated by individual rationality unless it coincides with self-interest.

Shubik's argument that it is inappropriate to ascribe preferences or wants to a group (see Shubik (1982, p. 124)) may be right in the context of exclusively individual self-interest. From the individual perspective, the players can have no concept of group preference; all they know about is their own preferences as expressed via their utility functions. Anything else is extra-game-theoretic. Consider Example 2.2 involving a factory scheduling problem. There is no way to assure that maximizing individual preferences maximizes corporate preferences. If we view the corporation as a group, we may succeed in defining group preferences, but we have no way to accommodate them, since we have no notion of group utility. We might consider viewing the corporation as a superplayer, but such an "individual" is not a decision maker and does not possess its own utility function. The bottom line is that group rationality simply cannot be defined in terms of individual rationality, and von Neumann–Morgenstern game theory was not designed for that purpose.

Although von Neumann–Morgenstern game theory does not accommodate the notion of group rationality, common sense certainly does. It is indeed possible for individual workers in a factory to not only want to do well themselves but also want each of their neighbors, as well as the entire corporation, to do well, even if the latter achievements come somewhat at their expense. It may be claimed that any such ulterior motives should be reflected in the individual utilities, but individual utilities are only a function of the options available to the other players, not of their preferences for those options. While it may be possible to modify the utility functions somewhat to account for social welfare, it is difficult to see how such a practice accommodates group preferences in a general way.

The strength of von Neumann–Morgenstern theory is that it produces solutions that are superior, according to individual rationality, to all alternatives. Its weakness is that individual rationality does not imply group rationality, nor vice versa. Its instrumentality is that it tells us about outcomes we can expect when individually rational decision makers negotiate. In this way it is a natural analysis tool. Its limitation is that it does not give procedures for actually doing the negotiating – it is not a natural synthesis tool. One can always artificially interweave negotiatory events into an explanatory story justifying how a decision should be obtained, even though the story is not an explicit part of the generative decision-making model and may be misleading. But synthesis requires that the decision makers must actually live the story and perform the negotiatory functions.

Decision problems involving only two players are seemingly simple. If the game involves conflict, then a Nash equilibrium seems to address the situation adequately, except for the annoying fact that such equilibria may not be unique. If the game is a coordination game, the Pareto solution concept makes sense, except that, here as well, Pareto equilibria may not be unique. Thus, even for the simplest non-trivial game situations, it is not obvious how to settle on a unique strategy. But things get much more difficult when we move to the general N-player game context. At least, when $N = 2$, either both the players form a single grand coalition or they each go their separate ways. When $N > 2$, there are many coalition possibilities, and none of them is binding. The dynamics of coalition formation is one of the largely unsolved problems of game theory. One reason it is unsolved is that solution concepts based on maximizing expectations are not constructive in the sense that, although they tell us how to recognize the optimal solution when we see it, they do not tell us how to find it. Although they may tell us what a good (perhaps even optimal) negotiated solution looks like, they do not tell us how to negotiate.

A number of maximal-expectations-based negotiatory solution concepts have been proposed. But remember that these solution concepts do not define processes; they are nothing more than justifications for processes that are presumed somehow to occur in the heat of battle. They do, however, lend themselves to the creation of story lines that sound convincing and provide reasonable explanations for behavior that intelligent self-interested decision makers would conceivably experience when negotiating.

Definition 7.1
A **cooperative game** is a game in which the players can originally choose a payoff together; that is, they have the opportunity to communicate and form binding agreements before the game is played.   □

It is axiomatic that players in a cooperative game will choose from the set of Pareto equilibria (see Definition 1.9); otherwise some players would sacrifice needlessly. A second principle is often invoked, namely the principle of security. Suppose a player were to "go it alone," and refused to cooperate. Such a player should never agree to a strategy vector (Pareto-optimal or not) unless its resulting individual payoff were at least as great as its security level (see Definition 1.12). This assumption is a necessary consequence of the basic axiom of the individual maximization of expectations and illustrates the restrictiveness of that point of view. It does not permit, for example, an expression of altruism, wherein a player may sacrifice some of its possible satisfaction for the benefit of others or for the benefit of the entire community.

Definition 7.2
The **negotiation set** is the set of all Pareto equilibrium strategy vectors whose component strategies meet the corresponding individual security levels.   □

The "go it alone" strategy is unnecessarily pessimistic in many situations, and rational players should be interested in exploring the possibilities that open up to them as a consequence of their cooperation with other players. In particular, when the payoff is such that it can be reapportioned among the players, the formation of coalitions can be extremely attractive.

Definition 7.3
Payoffs are said to be **transferable** if they are money-like, that is, they can be exchanged freely between players. These transfers are called **side-payments**.   □

Definition 7.4
A **coalition** is a group of players who have agreed to function as a unit. The **grand coalition** Γ is the coalition formed by all players in the game. The set of all possible coalitions is the **power set** $\mathcal{X}$, that is, the collection of all subsets of $\mathbf{X} = \{X_1, \ldots, X_N\}$.   □

Consider any coalition, $G \in \mathcal{X}$. Since we assume that payoffs are transferable, the payoff that G receives is well defined (e.g., the payoff may be money). Let the payoff function for X_i be denoted by π_i. (For additional details, see Appendix B.)

Definition 7.5
The **coalition payoff** for a coalition G is the mapping from the joint strategy space to the real numbers, $w_G \colon \mathcal{S} \to \mathbb{R}$, that characterizes the payoff that G receives as a result

of forming a coalition. Note that this function is not necessarily defined in terms of the individual payoff functions π_i. □

Now observe that G cannot fare worse than if all of the decision makers not in G combine against it in a single rival coalition. In other words, if the players were to coalesce into two groups, G and $\mathbf{X}\setminus G$, this would be the arrangement that could do the most damage to G.

Definition 7.6

The **characteristic function** of a coalition G is the maximum coalition payoff that G can be assured of achieving. Notationally, we define this value as follows. We first must decompose the joint strategy space into two subspaces, one composed of strategies for the members of G, the other for the members of $\mathbf{X}\setminus G$. Let these two subspaces be denoted $\mathcal{S}_G$ and $\mathcal{S}_{\mathbf{X}\setminus G}$, respectively. Then the characteristic value of the coalition G is

$$v(G) = \max_{s' \in \mathcal{S}_G} \min_{s'' \in \mathcal{S}_{\mathbf{X}\setminus G}} \{w_G(s', s'')\}.$$

The characteristic function is a generalization of the security level introduced earlier in the context of individual payoffs and reduces to that notion for single-member coalitions. □

An important property of the characteristic function is that it is superadditive.

Definition 7.7

A function v is said to be **superadditive** if, for any disjoint coalitions G_1 and G_2,

$$v(G_1 \cup G_2) \geq v(G_1) + v(G_2).$$ □

Superadditivity is a consequence of the fact that, acting together, the members of G_1 and G_2 can achieve everything that they could achieve if they acted in two separate subgroups, and they can possibly achieve more if they cooperate.

Definition 7.8

A game is said to be **essential** if there exist two coalitions such that they can strictly increase their joint security level by joining together; that is, if

$$v(G_1 \cup G_2) > v(G_1) + v(G_2)$$

for some coalitions G_1 and G_2. If this does not occur, that is, if forming coalitions does not ever provide improved security, then the game is said to be **inessential**. □

The characteristic function provides a general description of both individual and group interests under the von Neumann–Morgenstern tradition of game theory.

It provides a means of identifying coalitions that are justifiable according to the basic notions of rationality upon which the theory is built. There are three criteria that any such rational coalition must satisfy.

Let **s** be a candidate strategy for consideration as being rational. The first criterion is that, under the grand coalition, the entire group must receive at least its security level; that is, the sum of the individual payoffs under **s** must be at least as large as the characteristic function of the grand coalition:

$$\sum_{i=1}^{N} \pi_i(\mathbf{s}) \geq v(\Gamma).$$

But, by definition, $v(\Gamma)$ is not only achievable, but the most that is achievable by all decision makers working cooperatively, thus

$$\sum_{i=1}^{N} \pi_i(\mathbf{s}) \leq v(\Gamma).$$

These two relationships imply that

$$\sum_{i=1}^{N} \pi_i(\mathbf{s}) = v(\Gamma). \tag{7.1}$$

In other words, for a strategy to be acceptable it must be a Pareto equilibrium.

The second criterion is that, if **s** is implemented, no individual player should be required to receive less than its security level; that is,

$$\pi_i(\mathbf{s}) \geq v(X_i), \quad i = 1, \ldots, N. \tag{7.2}$$

Definition 7.9

Any strategy that satisfies both (7.1) and (7.2) is called an **imputation strategy** or **imputation**.[1]   □

The third criterion is motivated by the following observation. Suppose the players are entertaining an agreement to play strategy vector $\mathbf{s}'$, which, if consummated, would return the payoff vector $\{\pi_i(\mathbf{s}'), i = 1, \ldots, N\}$. Now let G be a potential coalition. If the sum of the individual payoffs received by the members of G when playing $\mathbf{s}'$ is less than the maximum coalitional payoff they can be guaranteed to receive if they were to form a coalition, then the members of G should, if rational, refuse to consummate $\mathbf{s}'$. For, by playing as a coalition, they can be assured of receiving at least $v(G)$, which may then be redistributed by side payments to make all members of the coalition

---

[1] In the literature, much of the discussion of this subject deals directly with the payoffs and not explicitly with strategies. This is satisfactory, since payoffs are functions of strategies and it is often simpler to operate in a payoff space. In this current development, however, we wish to operate in strategy space, so we deviate slightly from conventional practice.

Pareto better off than they would have been if $\mathbf{s}'$ were implemented. In the jargon of game theory, the coalition G should block $\mathbf{s}'$. Thus, the third rationality condition is that

$$\sum_{X_i \in G} \pi_i(\mathbf{s}') \geq v(G) \tag{7.3}$$

for every coalition G.

Definition 7.10
The **core** is the set of strategy vectors that satisfy all three rationality conditions: (7.1), (7.2), and (7.3). □

The core is a fundamental equilibrium concept. The strategy vectors that lie in it are those that survive an intense form of competition between players. Anyone may combine with anyone else to block a potential solution. There are no loyalties and no prior commitments. Nor are patterns of collusion prohibited. Until everyone is satisfied, negotiations may continue, and players are free to overturn any provisional arrangements.

But the core is no panacea. The requirements are so strenuous that, for many games, the core is empty. In fact, it can be shown (see, e.g., Bacharach (1976)) that, if a constant-sum game is essential, then the core is empty.

A famous example of a game with an empty core is the three-player game *Laissez-Faire*, which also goes by the name *Share a Dollar*. There are 100 pennies (a dollar) on the table, and it is decided by majority vote which coalition gets to share the dollar, which can be divided between two players only. A strategy is a three-tuple of the form $\mathbf{s} = \{s_1, s_2, s_3\}$, where s_i is the number of pennies that X_i receives. The payoff for X_i is defined as the number of pennies that comes to X_i; that is,

$$\pi_1(s_1, s_2, s_3) = s_1,$$
$$\pi_2(s_1, s_2, s_3) = s_2,$$
$$\pi_3(s_1, s_2, s_3) = s_3.$$

Since no player can completely eliminate the possibility of being frozen out of a deal, it is easy to see that the characteristic function for this game is

$$v(X_1) = v(X_2) = v(X_3) = 0,$$
$$v(X_1, X_2) = v(X_1, X_3) = V(X_2, X_3) = 100,$$
$$v(X_1, X_2, X_3) = 100.$$

The game is essential, since, for example, X_1 and X_2 may achieve a security level of 100 by cooperating, while the sum of their individual security levels is zero. Let us verify the claim that the core is empty for this game. Suppose it is not empty and that $\mathbf{s} = \{s_1, s_2, s_3\}$ belongs to it. Since, by hypothesis, elements of the core satisfy (7.3), this means that

$$\pi_1(s) + \pi_2(s) \geq 100,$$
$$\pi_1(s) + \pi_3(s) \geq 100,$$
$$\pi_2(s) + \pi_3(s) \geq 100.$$

Adding these three inequalities, we have

$$2\pi_1(s) + 2\pi_2(s) + 2\pi_3(s) \geq 300 \quad \text{or} \quad \pi_1(s) + \pi_2(s) + \pi_3(s) \geq 150,$$

but this is impossible, since $\pi_1(s) + \pi_2(s) + \pi_3(s) \leq 100$. Thus, the core is empty.

The dynamics of this game are vicious. Suppose that X_1 and X_2 provisionally agree on $s = \{50, 50, 0\}$. As long as X_1 and X_2 stick together, all is fine, but suppose X_3 now approaches, say, X_2 with a proposal of $s' = \{0, 60, 40\}$. This would surely entice X_2 away from s, since it would gain an extra 10 pennies. But nothing prohibits X_1 from approaching X_3 with an offer of the form $s'' = \{50, 0, 50\}$, which would entice X_3 away from a coalition with X_2. Such recontracting is endless. Every proposed arrangement can be frustrated by a counter-offer – every strategy vector is blocked. The game has no solution that will please every player, someone will be disgruntled. This is what it means for the core to be empty.

The empty core exposes the ultimate ramifications of a decision methodology based strictly on individual expectations maximization. This model imposes an insatiable, intemperate, and restless attitude on the players – hardly a constructive environment in which to conduct negotiations. Much research has focused on ways to extend the theory to validate solutions not in the core. Shapley (1953) suggests that players in an N-person game should formulate a measure of how much their joining a coalition contributes to its value and should use this metric to justify their decision to join or not to join. This formulation, however, requires the acceptance of additional axioms involving the play of composite games (Rapoport, 1970).

Another way to extend the notion of the core is to form coalitions on the basis of no player having a justifiable objection against any other member of the coalition, resulting in what is called the bargaining set. Also, it is certainly possible to invoke various voting or auctioning protocols to address this problem. Extra-game-theoretic considerations, such as friendship, habits, fairness, etc., may also be applied to modify the behavior of agents desiring to achieve some rationale for forming coalitions. All of these considerations, however, either require the imposition of additional assumptions or extend beyond consideration of strict individual rationality in rather *ad hoc* ways. Rapoport summarizes the situation for conventional N-person game theory succinctly:

If ... we wish to construct a *normative* theory, i.e., be in a position of advising "rational players" on what the outcomes of a game "ought" to be, we see that we cannot do this without further assumptions about what explicitly we mean by "rationality." (Rapoport, 1970, p. 136)

There are many variations to this basic theory, involving both cooperative and non-cooperative game theory (see, e.g., Hart and Mas-Colell (1994), but we will not pursue them. The core is an adequate example of the way one can generate rational explanations for why a strategy is a good one. The core does not provide us, however, with a constructive process by which such a solution can be actually reached – only how to recognize it when we see it. All "negotiations," in effect, take place before the game is played. Actually playing the game is anticlimactic. Perhaps this is what is supposed to happen. Maybe the possibilities are so rich and so varied that it is impossible to define a constructive procedure that would lead, in a sure-fire way, to a good solution. If so, then the role of conventional game theory as a mechanism for modeling negotiatory systems must be limited to telling us what to expect when rational decision makers negotiate. This is simply a fact of life under the expectations-maximizing paradigm. It leads to concepts that tell us what to do, but not how to do it. In particular, they cannot tell us how to build computer models of negotiation. They are instructive, but not constructive, and should not be asked to deliver more than they can.

Another stream of theory for the design of negotiatory systems is to rely more heavily on heuristics than on formal optimization procedures. The approach taken by Rosenschein and Zlotkin is to emphasize special compromise protocols involving pre-computed solutions to specific problems (Rosenschein and Zlotkin, 1994; Zlotkin and Rosenschein, 1996c, 1996b, 1996a). Formal models that describe the mental states of agents based upon representations of their beliefs, desires, intentions, and goals can be used for communicative agents (Cohen and Levesque, 1990; Cohen et al., 1990; Kraus and Lehmann, 1999; Lewis and Sycara, 1993; Shoham, 1993; Thomas et al., 1991; Wellman and Doyle, 1991). In particular, Sycara develops a negotiation model that accounts for human cognitive characteristics and views negotiation as an iterative process involving case-based learning and multi-attribute utilities (Sycara, 1990; Sycara, 1991). Kraus et al. (1998) provide logical argumentation models as an iterative process involving exchanges among agents to persuade each other and bring about a change of intentions. Zeng and Sycara (1997, 1998) develop a negotiation framework that employs a Bayesian belief-update learning process through which the agents update their beliefs about their opponent. Durfee and Lesser (1989) advance a notion of partial global planning for distributed problem solving in an environment of uncertainty regarding knowledge and abilities.

These approaches offer realistic ways to deal with the exigencies under which decisions must be made in the real world. However, they all have a common theme, which is that, if a decision maker could maximize its own private utility subject to the constraints imposed by other agents, it should do so. Although the preponderance of economic decision-making philosophy seems to be based ultimately upon the premise of individual rationality, it does not follow that this is the only viable model for the analysis of human systems. Nor, more importantly, does it follow that a model of behavior for human systems should be uncritically adopted for the design of artificial systems, particularly if the model is subject to criticism as *the* appropriate model of human behavior.

7.2   Satisficing negotiation

7.2.1   The negotiation theorem

Communities usually exist or are created for a purpose, and often the purpose is larger than the individual. When this is the case, the functionality of the community will depend on the interests of the group as well as of the individual. Since we cannot manufacture a notion of group rationality from individual rationality, we must abandon our demand for exclusive self-interest if we want to deal with group rationality. Fortunately, as was discussed in Section 6.4, a notion of group rationality may be expressed in terms of praxeic utility theory and the satisficing notion.

Definition 7.11
Let p_S and p_R be joint selectability and rejectability functions defined over the joint strategy space $\mathcal{S}$. The **satisficing strategy vector set at boldness** q is

$$\Sigma_q = \{\mathbf{s} \in \mathcal{S}: p_S(\mathbf{s}) \geq q p_R(\mathbf{s})\}. \qquad \square$$

The set Σ_q defines the set of group strategies that are satisficing for the entire community. The elements of this set are such that the overall avoidance of failure of the community, as expressed by p_S, outweighs the overall resource consumption of the community, as expressed by p_R. According to the principle of intrinsic rationality, all elements of Σ_q are good enough for the community. As discussed in Section 6.4, the interpretation of the concept of group satisfaction depends on the context. To reiterate, the notion of group preference is coordination/conflict neutral. For a competitive game, the group preference may be to oppose one another, while for a cooperative game the group preference may be to coordinate. Furthermore, the group preference can be emergent.

But what is good enough for the community may not coincide with what is good enough for the individuals. The individual satisficing sets

$$\Sigma_q^i = \{s_i \in \Sigma_i: p_{S_i}(s_i) \geq q p_{R_i}(s_i)\}$$

identify the individual strategies that are satisficing, or good enough, for the ith individual, and the satisficing rectangle

$$\mathfrak{R}_q = \Sigma_q^1 \times \cdots \times \Sigma_q^N$$

is the collection of individually satisficing sets. The two sets Σ_q and $\mathfrak{R}_q$ thus represent all strategies that are either good enough for the group or good enough for the individuals.

They correspond to well-defined notions of group and individual preferences. But, as we saw in Section 6.3, individual and joint decisions are not necessarily compatible, in that $\mathfrak{R}_q$ and Σ_q may be disjoint! There is, however, a more primitive and weaker notion of consistency that is guaranteed to hold.

Theorem 7.1

(The negotiation theorem.) If s_i is individually satisficing for X_i, that is, $s_i \in \Sigma_q^i$, then it must be the ith element of some jointly satisficing vector $\mathbf{s} \in \Sigma_q$.

PROOF

We will establish the contrapositive, namely, that if s_i is not the ith element of any $\mathbf{s} \in \Sigma_q$, then $s_i \notin \Sigma_q^i$. Without loss of generality, let $i = 1$. By hypothesis, $p_\mathbf{S}(s_1, \mathbf{v}) < q p_\mathbf{R}(s_1, \mathbf{v})$ for all $\mathbf{v} \in U_2 \times \cdots \times U_N$, so

$$p_{S_1}(s_1) = \sum_\mathbf{v} p_\mathbf{S}(s_1, \mathbf{v}) < q \sum_\mathbf{v} p_\mathbf{R}(s_1, \mathbf{v}) = q p_{R_1}(s_1), \text{ hence } s_1 \notin \Sigma_q^1. \qquad \square$$

Thus, if a strategy is individually satisficing, it is part of a jointly satisficing strategy vector, although it need not be part of all jointly satisficing strategy vectors. The converse, however, is not true: if s_i is the ith element of a jointly satisficing vector, it is not necessarily individually satisficing for X_i. (As a counterexample, consider the problem discussed in Section 6.3, and observe that F is the second element of a jointly satisficing set for the Lucy and Ricky problem for certain values of α, but F is not individually satisficing for Ricky for any values of α.)

The content of the negotiation theorem is that *no one is ever completely frozen out of a deal – every decision maker has, from its own perspective, a seat at the negotiating table.* This is perhaps the weakest condition under which negotiations are possible.

To ensure that negotiation is not vacuous, we must assume that the purposes of the individuals who are party to the negotiation are served only when a mutually acceptable agreement is reached. Thus, failure can be avoided only if the community agrees on a joint decision that is acceptable to all participants. Therefore, in addition to whatever collective goals that may exist, a group preference must include the desire to avoid failing to reach an agreement. This constraint suggests three principles of negotiation.

N-1: Negotiators are usually more concerned with meeting minimum requirements than with achieving maximum performance.

N-2: Negotiations should lead to decisions that are both good enough for the group (i.e., failure to agree is avoided) and good enough for each individual.

N-3: Negotiation is, institutionally, a narrowing down of options; it is more natural to work with a set of good enough joint options and invoke an iterative process to converge to a deal than to search directly for a single joint option that is a best compromise.[2]

---

[2] The concept of narrowing down does not mean that the set of possible options is closed and does not preclude the introduction of creative new options that can emerge as a result of exchanges that occur during negotiation.

Satisficing is well-suited to these principles. The joint satisficing set Σ_q and the individually satisficing sets Σ_q^i provide each X_i with information critical to negotiation: a complete assessment, from its point of view, of all joint options that are good enough for the group (at a minimum, avoiding failure to agree), and of all individual options that are good enough for itself (according to its own interests). The negotiation problem is for each X_i to arrive at a compromise of its own interests with the interests of others so as to avoid an impasse.

For a decision maker to compromise would require it to lower its standards of what is good enough. No such concept exists under the auspices of exclusive self-interest – optimization does not admit grades, or degrees. Satisficing, or being good enough, on the other hand, is a graded concept. For a decision maker to lower its standards, however, would require a good reason. One possible reason is for the good of the group, at least in the interest of avoiding of failure to reach a compromise.

A decision maker who possessed a modest degree of altruism would be willing to undergo some degree of self-sacrifice in the interest of others. Such a decision maker may be viewed as an **enlightened liberal**; that is, one who is intent upon pursuing its own self-interest but gives some deference to the interests of others. Such a decision maker would be willing to lower its standards, at least somewhat and in a controlled way, if doing so would be of benefit to others.

Even in a non-harmonious negotiation scenario where altruism is not a factor, a decision maker may be inclined to lower its standards in order to avoid failure, especially if the consequences of the group failing to achieve a mutually agreeable decision are high (and thus all players are frustrated) compared with the individual compromising its individual interests.

The natural way for X_i to express a lowering of its standards is to decrease its boldness. Generally, we may set $q = 1$ to reflect equal weighting of the desire to avoid failure and the desire to conserve resources. By decreasing q, we lower the standard for failure avoidance relative to resource conservation and thereby increase the size of the satisficing set. As $q \to 0$ the standard is lowered to nothing, and eventually every option is satisficing for X_i. Consequently, if all decision makers are willing to reduce their standards sufficiently, a compromise can be achieved.

Definition 7.12
Let q_i be the boldness level for decision maker X_i. The **boldness vector** is the array $\mathbf{q} = (q_1, \ldots, q_N)$. The **least bold value** is $q_L = \min\{q_1, \ldots, q_N\}$. □

Definition 7.13
The set of all strategy vectors that are also individually satisficing for X_i is X_i's **compromise set** $\mathbf{C}_i$, given by

$$\mathbf{C}_i = \{\mathbf{s} = \{s_1, \ldots, s_N\} \in \Sigma_{q_L} : s_i \in \Sigma_{q_i}\}.$$

□

Note that the standard for a compromise is set by the least bold value, q_L, which reflects the notion that the standards of a group can be no higher than the standards of any member of the group.

Theorem 7.2
$C_i \neq \emptyset, i = 1, \ldots, N.$

PROOF

The negotiation theorem implies that the set

$$D_i = \{s = \{s_1, \ldots, s_N\} \in \Sigma_{q_i} : s_i \in \Sigma_{q_i}\} \neq \emptyset.$$

But $D_i \in \Sigma_{q_i}$, and since $q_L \leq q_i$, we have $\Sigma_{q_i} \subset \Sigma_{q_L}$, hence $D_i \in C_i$. □

Definition 7.14
A strategy vector $s = (s_1, \ldots, s_N)$ is a **satisficing imputation at boldness q** if $p_S(s) \geq q_L p_R(s)$ and $p_{S_i}(s_i) \geq q_i p_{R_i}(s_i)$ for $i = 1, \ldots, N$. That is, the strategy vector is jointly satisficing for the group and each component option is individually satisficing for the corresponding member of the group.

The **satisficing imputation set at q**, denoted N_q, is the set of satisficing imputations at q, and is given by $N_q = \cap_{i=1}^{N} C_i$. □

Definition 7.15
A **rational compromise at q**, designated s^*, is a satisficing imputation that maximizes the joint selectability to rejectability ratio; that is,

$$s^* = \arg\max_{s \in N_q} \frac{p_{S_1 \cdots S_N}(s)}{p_{R_1 \cdots R_N}(s)}.$$ □

A rational compromise provides maximal benefit for the group while ensuring that each decision maker's preferences are not compromised more than the decision maker permits. If $N_q = \emptyset$, then no rational compromise is possible at boldness q. Thus, at least some of the agents must be willing to lower their standards if a compromise is to occur. It is reasonable to assume that there is a lower limit to how much any decision maker would be willing to compromise. If that limit is reached for some decision maker and no agreement is reached, then that decision maker must break off negotiations and declare an impasse. Until that point is reached, however, negotiations may proceed in good faith according to the algorithm given in Figure 7.1, which is suitable for communities of enlightened liberals.

Reducing the boldness q_i is a controlled way to relax the standards of rationality, which may be necessary in difficult situations if a compromise is to be reached. The amount that q_i must be reduced below unity is a measure of the amount of compromising needed to reach a mutually acceptable solution. The satisficing imputation set is similar

Step 1: X_i forms $\Sigma_{q_L}^i$ and $\Sigma_{q_i}^i$, $i = 1, \ldots, N$; initialize with $q_i = 1$.

Step 2: X_i forms its compromise set by eliminating all strategy vectors for which its component is not individually satisficing, resulting in $\mathbf{C}_i = \{\mathbf{s} \in \Sigma_{q_L}^i : s_i \in \Sigma_{q_i}^i\}$.

Step 3: Broadcast $\mathbf{C}_i$ and q_i to all other participants, receiving similar information from them.

Step 4: Form the satisficing imputation set, $\mathbf{N_q} = \cap_{j=1}^N \mathbf{C}_j$. If $\mathbf{N_q} = \emptyset$, then decrement q_j, $j = 1, \ldots, N$, and repeat previous steps until $\mathbf{N_q} \neq \emptyset$.

Step 5: X_i implements the ith component of the rational compromise

$$\mathbf{s}^* = \arg\max_{\mathbf{s} \in \mathbf{N_q}} \frac{p_{S_1 \ldots S_N}(\mathbf{s})}{p_{R_1 \ldots R_N}(\mathbf{s})}.$$

Figure 7.1: The Enlightened Liberals negotiation algorithm.

in general structure to the von Neumann–Morgenstern imputation set (i.e., the set of payoffs that are both jointly and individually secure (Shubik, 1982)). But it differs significantly in that the superlative notion of security (maximin value) is replaced with the dichotomous notion of satisficing. This leads to a theory of social behavior that is very different from standard N-person von Neumann–Morgenstern game theory.

7.2.2   The Resource Sharing game

The following simple example illustrates the fundamental differences between substantive and intrinsic rationality. Suppose a factory operates N processing sectors that function independently of each other, except that, if their power requirements exceed a fixed threshold, they must draw auxiliary power from a common source. Unfortunately, there are only $N - 1$ taps to this auxiliary source, so one of the sectors must operate without that extra benefit. Although each sector is interested in its individual welfare, it is also interested in the overall welfare of the factory and is not opposed to making a reasonable compromise in the interest of overall corporate success.

Let U denote the set of auxiliary power levels that are feasible for each X_i to tap and let $f_i: U \to [0, \infty)$ be an objective function for X_i; that is, the larger f_i, the more effectively X_i achieves its goal. X_i's choice is tempered, however, by the total cost of power, as governed by an anti-objective function, $g_i: U \to [0, \infty)$, such that, the smaller g_i, the less the cost. Work cannot begin until all players agree on a way to apportion the auxiliary power. Table 7.1 displays these quantities for a situation involving three decision makers.

A standard approach under substantive rationality is to view this as a cooperative game. The payoffs may be obtained by combining the two objective functions, yielding

Table 7.1: The objective functions for the
Resource Sharing game

U	f_1	g_1	f_2	g_2	f_3	g_3
0.0	0.50	1.0	0.10	1.0	0.25	1.0
1.0	2.00	2.0	2.00	3.0	0.50	5.0
2.0	3.00	4.0	3.00	6.0	1.00	5.0
3.0	4.00	5.0	4.00	9.0	2.00	5.0

individual payoff functions of the form

$$\pi_i(u_i) = \alpha_i f_i(u_i) - \beta_i g_i(u_i) \tag{7.4}$$

$i = 1, 2, 3$, where α_i and β_i are chosen to ensure compatible units. To achieve this
compatibility, we normalize f_i and g_i to unity by setting

$$\alpha_i = \frac{1}{\sum_{u \in U} f_i(u)},$$

$$\beta_i = \frac{1}{\sum_{u \in U} g_i(u)}.$$

A classical way to solve this problem is to invoke a negotiation protocol. Of the
various protocols that are possible, the only one that does not require assumptions
additional to that of self-interested expectations maximization is the core. Recall that
the core is the set of all payoffs that (i) are Pareto equilibria, (ii) ensure that each
decision maker achieves its individual security level, and (iii) ensure that every coalition
achieves its group security level. Unfortunately, as with *Laissez-Faire*, the core is empty.
Essentially, this is because only two decision makers can share in the auxiliary power
source, effectively disenfranchising the third decision maker. This situation potentially
leads to an unending round of recontracting, where participants continually make offers
and counter-offers in a fruitless attempt for all to maximize their expectations.

Rather than apply yet another patch to substantive rationality, we develop a solution
in the new garb of intrinsic rationality by viewing the decision makers in their true
character as enlightened liberals who are willing to accept solutions that are serviceably
good enough for both the group and the individuals. From the point of view of the group,
an option is satisficing when the joint selectability exceeds the joint rejectability. Let
us define joint rejectability as the normalized product of the individual costs functions,
namely,

$$p_{R_1 R_2 R_3}(u_1, u_2, u_3) \propto g_1(u_1)g_2(u_2)g_3(u_3),$$

where "$\propto$" means the function has been normalized to sum to unity. To compute the
joint selectability, we note that, under the constraints of the problem, only two of the

agents may use the auxiliary power source. We may express this constraint by defining the joint selectability function as

$$p_{S_1 S_2 S_3}(u_1, u_2, u_3) \propto \begin{cases} f_1(u_1) f_2(u_2) f_3(u_3) & \text{if } \mathbf{u} \in \Pi, \\ 0 & \text{otherwise,} \end{cases}$$

where Π is the set of all triples $\mathbf{u} = \{u_1, u_2, u_3\}$ such that exactly one of the entries is zero. The individual rejectability and selectability marginal mass functions are obtained by summing over these joint mass functions according to 6.17 and 6.18.

The enlightened liberals algorithm yields, for $q > 0.88$, an empty satisficing imputation set. But, when q is decremented to 0.88, the satisficing imputation set is

$$\mathbf{N} = \{\{0, 1, 3\}, \{0, 2, 3\}, \{0, 3, 3\}\}$$

and the rational compromise is $\mathbf{u}^* = \{u_1^*, u_2^*, u_3^*\} = \{0, 1, 3\}$ which, coincidentally, is the Pareto optimal solution. It is not surprising that, at unity boldness, there are no options that are simultaneously jointly and individually satisficing for all participants, since there is a conflict of interest (recall that the core is empty). But, if each individual adopts the point of view offered by intrinsic rationality, it gradually lowers its personal standards to a point where it is willing to be content with reduced benefit, provided its costs are reduced commensurately, in the interest of the group achieving a collective goal. The amount q must be reduced to reach a jointly satisficing solution is an indication of the difficulty experienced by the participants as they attempt to resolve their conflicts. Reducing boldness is a gradual mechanism by which decision makers subordinate individual interest to group interest. In other words, individual interests are eventually subordinated so that no individual interests conflict. This mechanism is very natural in the regime of making acceptable tradeoffs, but is quite foreign to the concept of maximizing expectations ("finally, something we can agree to" versus "nothing but the best").

The attitude parameters for this decision problem are given in Table 7.2. We interpret these values as follows.

1. Group diversity is high and group tension is low, indicating that, as a group, the system is fairly well suited for its environment. This is because overall productivity increases with the tapping of the additional power, even though some of the agents benefit more than others. This interpretation is borne out by fact that the group is gratified with the joint decision.

2. X_2 has the lowest diversity and the highest tension. This situation is reflected in the structure of $\mathcal{C}$, since X_2 has several choices that are good enough, but is either dubious or ambivalent about all of them. Thus, X_2 experiences the most conflict in making decisions, with a tension value of 1.307 with an upper bound of 1.89. X_2 is dubious about its component of the rational compromise, since, although the performance to cost ratio is greater than 1, both are below the level obtainable with a uniform distribution of selectability and rejectability. Thus, although X_2 does

Table 7.2: Attitude parameters for the
Resource Sharing game ($q = 0.88$)

Agent	$H(p_S)$	$H(p_R)$	Diversity	Tension
X_1	1.837	1.784	0.550	0.931
X_2	1.811	1.680	0.025	1.307
X_3	1.665	1.823	1.214	0.733
Group	4.300	5.282	2.851	0.505

Agent	$p_S(u_i^*)$	$p_R(u_i^*)$	Attitude
X_1	0.402	0.083	gratification
X_2	0.204	0.158	dubiety
X_3	0.276	0.312	ambivalence
Group	0.051	0.004	gratification

Table 7.3: Cost functional values for the Resource Sharing
game

Agent	$\pi(0)$	$\pi(1)$	$\pi(2)$	$\pi(3)$	Dynamic range
X_1	−0.031	0.044	−0.018	0.004	0.075
X_2	−0.042	0.062	0.014	−0.034	0.104
X_3	0.004	−0.179	−0.046	0.221	0.400

not pay a great deal, it does not get a great benefit in return. X_2 experiences only mediocre performance; this is the reason for its relatively high stress – it is not very well suited for its environment. This assessment may be used by X_2 to change its relationship with its environment. For example, it may attempt to restructure its costs or reconfigure its processing capabilities.

3. X_3 is ambivalent, meaning that, even though it achieves a good benefit, its costs are commensurately high. It is interesting to note that X_3 is the only one who must actually compromise its standards in order to achieve a group solution; this is because X_3 has no cheap options.

4. A perhaps surprising feature of this example is that X_1, although frozen out of the coalition, is the least conflicted of the three agents; in fact, it is gratified with the joint solution. We may gain some insight as to why this is so by examination of the cost functional, π, as illustrated in Table 7.3. We observe from this table that X_1 has the smallest dynamic range of cost. Thus, X_1 is less insensitive to the outcome than is X_2 or X_3. Since X_1 has little to gain or lose, it is not seriously conflicted with its neighbors and has little incentive to compete for resources.

As this example illustrates, indicators of attitude are not indicators of performance. Rather, they are indicators of how well suited, or tuned, the decision maker is to function in its environment. A decision maker in a highly stressful environment may in fact be making very good decisions. High stress, however, can motivate the decision maker to re-evaluate its ecological standing.

7.2.3   Intrinsic decisions

A characteristic of classical negotiation concepts is that the notion of what is rational is externally imposed upon the decision makers. They are obligated to comply with the strict dictates of individual rationality and are not at liberty to compromise in any way that would grant an advantage to anyone at their expense. This is a significant handicap when negotiating. The essence of true negotiating is the ability and authority to make tradeoffs. Without such authority, a decision maker is constrained to be rigid and unyielding. The problem with compromise under substantive rationality is that making concessions with respect to individual self-interest places the decision maker in the untenable position of having to abandon all of its standards completely, which it cannot do and maintain any semblance of individual rationality.

Under the satisficing paradigm, however, the standards are not rigid. Rather, there are graded degrees of what is good enough, as determined by the boldness parameter q. The criterion for evaluating options is internal; rationality is intrinsic. The decision maker has the ability to judge each option on its own merits, rather than relying on performance relative to other options. Thus, a decision maker can exhibit some flexibility. It can accommodate the notion of compromise without abandoning its standards completely.

A significant difference between the satisficing imputation set and the core is that we do not need to compute jointly satisficing sets for all possible coalitions. This is because coalitions are artifacts of substantive rationality. The motivation for coalitions in cooperative game theory is the need to reduce a group decision-problem to a super-individual decision problem so that individual expectations maximization can be applied to decide which coalition should be instantiated. This need is obviated under intrinsic rationality. Cooperative associations, if they form at all, emerge implicitly from the structure of the interdependence function. For this resource sharing problem, coalitions do not naturally exist, and they should not be fabricated simply to obtain a plausible solution.

Substantive rationality – choosing the best and only the best – is perhaps the strongest possible notion of rationality, but, by the argument presented here, it is more important to be good enough than to be best. Intrinsic rationality – getting something that is acceptable to everyone – is a weaker notion, but it is also more fundamental. Evaluating dichotomies is a more primitive activity than searching for extrema. It is local, rather than global; it is internal to the issue, rather than external.

Satisficing theory represents a principled alternative between strict optimality and pure heuristics and mitigates the problems that arise with both extremes. The main problems with optimization-based negotiation are that (a) the criteria do not address possible conflicts between group and individual preferences; and (b) optimization is not constructive – it may identify a best compromise, but does not provide a procedure for reaching it. Purely procedurally rational approaches to negotiation address both of these problems, but they lack the demonstrated capacity for self-policing. By softening but not abandoning the rigid demands of substantive rationality, intrinsic rationality retains the capacity for self-policing and provides a framework for the construction of negotiatory processes that account for both group and individual preferences.

7.3 Social welfare

Social choice theory is the study of the relationships between the preferences of a society and the preferences of its members. It differs from general game theory in one somewhat subtle way. With games, each individual participant considers its own set of options, which may be different for each participant. Social choice theory, on the other hand, usually deals with voting situations, where all individuals share the same option set, and one and only one option is chosen as the winner. The social choice problem is to reconcile collective choice with individual choice. Our discussion begins with a review of the fundamental result of any social choice theory that is built on the concept of total orderings. We then develop the concept of satisficing social welfare.

7.3.1 Arrowian social welfare

Suppose that every member of a collection of decision makers has its own total ordering function over a set of propositions. When is it possible to devise a voting scheme as a function of the individual preference orderings that provides a total ordering of the preferences of the entire group of decision makers? In other words, under what conditions can there exist a decision rule that preserves reflexivity, antisymmetry, transitivity, and linearity for the society? Such a function is called an **Arrowian social-welfare function**. Arrow's impossibility theorem provides an answer to this question. He proceeds by first postulating a set of properties of democratic choice that would be reasonable for the members of any society to desire. Sen's account (Sen, 1979) of this result lists four properties.

R: There must be unrestricted freedom of individual choice. Each individual is allowed to choose any of the possible orderings of the propositions.

P: The Pareto principle must apply. That is, if every member of the society prefers u_i to u_j, then the society must also prefer u_i to u_j.

I: The social choice rule must be independent of irrelevant alternatives. That is, the social choice over a set of propositions must depend on the orderings of the individuals only over those propositions, and not the preferences with respect to propositions that are not members of the set.

D: There must not be a dictator. That is, there must not be an individual such that, whenever it prefers u_i to u_j, then the society must also prefer u_i to u_j, irrespective of the preferences of the other individuals.

Arrow proved, with his famous impossibility theorem, that if the individuals each possess total orderings over the preference set, then there is no method of combining individual preferences to generate an Arrowian social-welfare function that satisfies properties R, P, I, and D (Arrow, 1951; Sen, 1979).

An important application of this theorem is to establish the fact that any collective choice function that honors the four conditions and is reflexive, antisymmetric, and linear must necessarily be intransitive. This is the basis for the Voters' Paradox that is often used to illustrate the complexity of multi-agent decision making. It is illustrated in the following example, evidently discovered by Nanson (1882).

Example 7.1 Voters' Paradox. Consider a society of three individuals, X_1, X_2, and X_3, and three alternatives, u_1, u_2, and u_3. The preferences of these three individuals are

X_1: $u_1 \succeq u_2$ and $u_2 \succeq u_3$,

X_2: $u_2 \succeq u_3$ and $u_3 \succeq u_1$,   (7.5)

X_3: $u_3 \succeq u_1$ and $u_1 \succeq u_2$.

If we adopt the method of majority voting, we see that u_1 defeats u_2 by two votes to one and u_2 defeats u_3 by the same margin, so transitivity requires that u_1 should defeat u_3. But, in fact, u_3 defeats u_1 by two votes to one – thus the paradox.

7.3.2   Satisficing social welfare

Conventional approaches to social welfare start by first considering individual orderings, with each individual defining its preference rankings independently of explicit consideration of possible interdependencies on others (Sen, 1979). Social preferences are then considered as functions of the individual preferences. But, as we see, it can be impossible to construct a social-welfare function in a way that avoids problems such as intransitivity if we also insist on other reasonable and desirable characteristics.

The root of the problem is the assumption that individual preferences can be isolated and formed independently of the interests of others. In actuality, individual preferences may not be independent, and this observation is the basis for attempts to identify conditions that eliminate the possibility of paradoxes. Sen (1979), for example, has shown that group intransitivity can be avoided if the set of voter preferences are "value restricted" in that all voters agree that some alternative is not best, or some alternative is not intermediate, or some alternative is not worst, in anyone's preference ranking. Thus, one way to avoid paradoxes is somehow to get behind the scene and impose certain constraining dependencies between the voters' preferences.

While the observation that interdependencies between individual voters' preferences may be useful for constructing hedges against paradoxes, it is also, and more fundamentally, a recognition of the fact that such interdependencies can exist, and when they do, they should not be ignored. Unfortunately, however, if we view individual preferences as the starting point for the construction of a social-welfare function, there is no systematic way to account for these preference interdependencies. Perhaps, if we start at the headwaters of preference formulation, rather than somewhere downstream, we may be able to provide more comprehensively and systematically for these interdependencies. This approach will require us to implement a mechanism to express the natural preference couplings, should there be any, between individuals.

One way to do this is to move from the restricted domain of interest localization that is associated with the superlative paradigm of global ordering to the more universal view of interest globalization that is associated with the comparative paradigm of local ordering. The interdependence function provides a convenient device with which to account for the connections between preferences. Since this function has the structure of a probability mass function, it permits the modeling of correlations between decision-maker preferences, should they exist.

Let $\{X_1, \ldots, X_N\}$ be a set of voters and let $U = \{u_1, \ldots, u_n\}$ be a set of alternatives, one of which will ultimately be chosen as a result of a collective decision of all voters. Let $\mathbf{U} = U_1 \times \cdots \times U_N$ denote the product set consisting of all possible N-tuples of voters' choices. Let $\mathbf{v}$ and $\mathbf{w}$ denote arbitrary elements of $\mathbf{U}$; that is, $\mathbf{v} = \{v_1, \ldots, v_N\}$, where $v_i \in U$, $i = 1, \ldots, N$, is a vector of individual choices; similarly for $\mathbf{w}$. The interdependence function is a mass function $p_{SR} \colon \mathbf{U} \times \mathbf{U} \to [0, 1]$ that expresses all of the positive and negative dependencies that exist between the voters' preferences.

From the interdependence function we may extract the joint selectability and rejectability functions by integrating out the cross-interdependencies between selectability and rejectability, yielding

$$p_S(\mathbf{v}) = \sum_{\mathbf{w} \in \mathbf{U}} p_{SR}(\mathbf{v}; \mathbf{w}),$$

$$p_R(\mathbf{v}) = \sum_{\mathbf{w} \in \mathbf{U}} p_{SR}(\mathbf{w}; \mathbf{v}).$$

Although these functions represent the joint selectability and rejectability of all possible voting combinations, in the end only one of the alternatives will be implemented, and the benefit and cost to the society as a whole for each alternative may be found by evaluating these functions at the vectors $\{u_i, \ldots, u_i\}$ for $i = 1, \ldots, n$.

Definition 7.16

The **satisficing social-welfare function** is given by

$$W(u) = p_S(u, \ldots, u) - p_R(u, \ldots, u). \tag{7.6}$$

An option that maximizes the satisficing social-welfare function is a **collective choice**, denoted by

$$u_C = \arg\max_{u \in U} W(u). \tag{7.7}$$

$\square$

The satisficing preferences for each individual are obtained by first computing the marginal selectability and rejectability functions, as given by 6.17 and 6.18 and then computing the satisficing sets for each, as given by 6.19.

Definition 7.17
The **consensus set** is the intersection of all individual satisficing sets.

$$C_q = \cap_{i=1}^{N} \Sigma_q^i. \tag{7.8}$$

If $u_C \in C_q$, then u_C is an **acclamation**.

$\square$

An element of the consensus set is good enough for every individual, and an acclamation is not only good enough for every individual, but is also in the best interest of the society; that is, it maximizes the satisficing social welfare function. If $C_q \neq \emptyset$ but there are no acclamations, then society as a whole may not be well served by the outcome, even though the outcome is good enough for each individual. If the consensus set is empty, then there are no mutually acceptable alternatives and no outcome can be pleasing to all individuals.

Observe that, unlike the social-welfare functions proposed by Arrow and others under the paradigm of substantive rationality, the satisficing social-welfare function is not a function of individual preferences, except in the special case of complete voter inter-independence, that is, if no individual's preferences are influenced by any other individual's preferences. In this case, the interdependence function is composed of the product of marginal selectability and rejectability functions, that is,

$$p_{SR}(v_1, \ldots, v_N; w_1, \ldots w_N) = \prod_{j=1}^{N} p_{S_j}(v_j) \prod_{j=1}^{N} p_{R_j}(w_j),$$

where p_{S_j} and p_{R_j} are the marginal selectability and rejectability mass functions for X_j. The satisficing social-welfare function is then

$$W(u) = \prod_{j=1}^{N} p_{S_j}(u) - \prod_{j=1}^{N} p_{R_j}(u). \tag{7.9}$$

Let us now examine the Voters' Paradox from the satisficing point of view and assume complete inter-independence. To do so, we must define the selectability and rejectability functions, which requires us to provide operational definitions for failure

Table 7.4: Selectability and rejectability for the
Voters' Paradox under conditions of complete
voter inter-independence

U	p_{S_1}	p_{S_2}	p_{S_3}	p_{R_1}	p_{R_2}	p_{R_3}
u_1	0.500	0.167	0.333	0.333	0.333	0.333
u_2	0.333	0.500	0.167	0.333	0.333	0.333
u_3	0.167	0.333	0.500	0.333	0.333	0.333

Table 7.5: Conditional selectability for the
correlated Voters' Paradox

u	$p_{S_3|S_2}(u|u_1)$	$p_{S_3|S_2}(u|u_2)$	$p_{S_3|S_2}(u|u_3)$
u_1	0.100	0.000	0.333
u_2	0.000	0.100	0.167
u_3	0.900	0.900	0.500

avoidance and resource conservation. Perhaps the simplest way to frame this problem
is to associate the ordinal preference orderings given by (7.5) with selectability (failure
avoidance) and to view resource conservation as simply the cost of casting a vote, with
the same cost to each alternative. Using a simple numerical scale of 3 being best, 2
next best, and 1 worst for selectability, and modeling rejectability with the uniform
distribution, the selectability and rejectability functions become, upon normalization,
those shown in Table 7.4, from which we compute the individual satisficing sets as

$$\Sigma_q^1 = \{u_1, u_2\},$$
$$\Sigma_q^2 = \{u_2, u_3\},$$
$$\Sigma_q^3 = \{u_1, u_3\}.$$

We immediately see that the consensus set is empty. We also see that the satisficing
social-welfare function is constant. We may conclude that, under a condition of inter-
independence, the satisficing solution sheds no additional light on the Voters' Paradox.

We now consider a way to enrich this example by allowing less than complete
inter-independence between voters' preferences. Let us suppose that X_1 and X_2 are
preference-independent, but that the preferences of X_3 are coupled to those of X_2. The
joint selectability function may then be factored as

$$p_{S_1 S_2 S_3}(v_1, v_2, v_3) = p_{S_1}(v_1) p_{S_2}(v_2) p_{S_3|S_2}(v_3|v_2),$$

where p_{S_1} and p_{S_2} are as given in Table 7.4, but $p_{S_3|S_2}$ is given by Table 7.5.

Straightforward calculation yields the marginal selectability functions that are presented in Table 7.6. The marginals for X_1 and X_2 are unchanged but, although X_3's

Table 7.6: Marginal selectability and rejectability for the correlated Voters' Paradox

U	p_{S_1}	p_{S_2}	p_{S_3}	p_{R_1}	p_{R_2}	p_{R_3}
u_1	0.500	0.167	0.128	0.333	0.333	0.333
u_2	0.333	0.500	0.106	0.333	0.333	0.333
u_3	0.167	0.333	0.766	0.333	0.333	0.333

marginals have changed, X_3's ordinal preferences remain as

X_3: $u_3 \succeq u_1$ and $u_1 \succeq u_2$.

The individual satisficing sets, for $q = 1$, are

$\Sigma_q^1 = \{u_1, u_2\},$

$\Sigma_q^2 = \{u_2, u_3\},$

$\Sigma_q^3 = \{u_3\}.$

The consensus set is empty, as with the original formulation.

The satisficing social-welfare function for this last case, however, is not constant, but is easily computed from (7.6) to be

$W(u_1) = -0.033,$

$W(u_2) = -0.020,$

$W(u_3) = -0.009.$

There is now a unique collective choice, namely, u_3, even though the individual preferences have the same ordinal ranking. Note that the paradox is preserved when considering individual preferences only, because the interdependencies between the voters are not revealed by the marginal selectability functions. Clearly, the individual preferences do not tell the whole story when conditional preferences exist.

This phenomenon is exactly the same thing as happens with correlated random variables in statistical inference theory. One can derive the marginal statistical distributions from the joint distribution, but the joint distribution cannot be obtained from the marginals unless the random variables are statistically independent. The joint distribution is the more basic quantity and contains more information than the marginals. In our context, the joint selectability function contains information that cannot be derived from the individual selectability functions.

The major difference between social welfare *à la* Arrowian analysis and social welfare *à la* the interdependence function is that, under the former paradigm, individual preferences are basic, and social welfare is viewed as a function of individual preferences. Under the latter paradigm, individual preferences may be dependent upon preference relationships involving others. The issue is really one of *interest localization* versus *interest globalization*, as introduced in Section 2.1.

While it is debatable which of these paradigms is a better model for human behavior, the major interest in this book is to develop decision logic formalisms for artificial societies. It is likely that the choice of paradigm will depend largely on the purpose for which the society is to be developed. For conflictive or competitive applications, it may be that the interest-localization paradigm will be more effective but, for cooperative environments, it is likely that interest globalization will be more appropriate.

8 Complexity

> Complexity is no argument against a theoretical approach if the complexity arises not out of the theory itself but out of the material which any theory ought to handle.
> Frank Palmer
> *Grammar* (Penguin, 1971)

Uncertainty and complexity are opposite sides of the nescience coin. One who suffers from uncertainty is frustrated by a lack of knowledge; one who suffers from complexity is frustrated by a lack of know-how. Knowing what to do is one thing, but knowing how to do it is quite another thing. Even if one knows of a way, in principle, to solve a problem, the computational burden of the solution may be impossible with existing technology. In such cases the solution must await the development of improved methods of computation.

Apparent complexity can often be reduced by eliminating irrelevant attributes of the decision problem, a procedure that Rasmusen terms "no-fat modeling." This approach is as follows: "First, a broad and important problem is introduced. Second, it is reduced to a very special but tractable model that hopes to capture its essence. Finally, in the most perilous part of the process, the results are expanded to apply to the original problem" (Rasmusen, 1989). Once the problem has been reduced to its essence, game theory provides a way to compute solutions to be expanded to the original problem.

Perhaps the most important simplifying assumption that is employed when reducing a problem to a form tractable for von Neumann–Morgenstern game theory is individual rationality. This assumption motivates another, somewhat subtle, assumption: social relationships (i.e., relationships that may exist between the players, such as emotions, attitudes, etc.) are excluded from consideration when defining the expected utilities of the players. This exclusion is entirely appropriate under the paradigm of individual rationality, since the players are assumed to be self-interested optimizers who have no concern for the interests of others. Such players are asocial. For example, with the von Neumann–Morgenstern treatment of the Prisoner's Dilemma, jail-time is assumed to be the only issue that is relevant to an individual's decision and hence is the only quantity represented in the payoff matrix. There is no attempt to accommodate any relationships that may exist between them, and an attempt to do so would be seen by many game theorists as a fundamental change that would transform the Prisoner's Dilemma into an entirely different game. With von Neumann–Morgenstern game theory, the payoff array *is* the game; once it is defined, the story line is irrelevant.

Asocial behavior is not entirely excluded from consideration by game theory, but the only way social relationships can be accommodated is for a player categorically to modify its own utility with the utilities of others. But, as I have claimed earlier, this only simulates social relationships such as altruism and does not fundamentally alter the structure of the game. In such a case it is only the *definition* of self-interest that changes, not the *criterion* for defining preferences. As Sen (1990, p. 19) observed: "It is possible to define a person's interests in such a way that no matter what he does he can be seen to be furthering his own interests in every isolated act of choice . . . no matter whether you are a single-minded egoist or a raving altruist or a class-conscious militant, you will appear to be maximizing your own utility in this enchanted world of definitions."

The paradoxes and dilemmas that frequently arise in game theory may be charming and even entertaining[1] but, if such apparent incongruities arise due to the elimination of critical social relationships from the game model that are only *post factum* re-inserted to interpret the phenomena as incongruities, they cannot serve as valid models of the problem under consideration. Removing the "fat" from a model must not reduce it to skin and bones. When the opportunity exists to coordinate for the benefit of others, eliminating the sociological issues that could engender such cooperation from the individual expected utility functions can compromise the essence of the game and make it difficult or impossible to comply with the third of Rasmusen's steps, namely, to expand the results to the original problem. Confining the players solely to the consideration of their own interests may make a problem more tractable, but it may also make it unnecessarily more difficult to account for social relationships.

Although humans may be able to make the judgments necessary to expand an overly simplified model to the context of the original problem, it is less likely that artificial decision makers will be able to make such inferences. Consideration of social relationships may make the model more complex, but if additional complexity is required to account for all relevant factors, it cannot be safely eliminated. Since satisficing game theory provides a mechanism for accounting for social relationships, it may offer a more accommodating model than does conventional game theory when the players are not appropriately characterized as asocial, optimization-obsessed individuals.

The price to be paid for an enlarged view is increased complexity. As discussed in Chapter 6, the ability to define conditional preferences provides the opportunity to account for social relationships as well as individual interests. The resulting interdependence function is more complicated in its structure than von Neumann–Morgenstern expected utility functions because it accounts for the interaction of two kinds of

---

[1] Two delightful as well as serious and instructive books on game theory are S. J. Brams' *Superior Beings: If They Exist, How Would We Know?* (New York: Springer-Verlag, 1983) and J. D. Williams' *The Compleat Strategyst* (New York: Dover, 1986). These books apply game theory to numerous social situations and discuss a number of paradoxes.

preferences among decision makers, namely, the avoidance of failure and the conservation of resources. One of the potential advantages of satisficing games, however, is that the interdependence function can be constructed from local conditional information and therefore the need for exhaustive global preference orderings can sometimes be eliminated. Consequently, its specification may actually be simpler than the specification of von Neumann–Morgenstern utilities, since global preferences may emerge from local preferences via the conditioning structure.

Nevertheless, satisficing game theory is potentially more complex than von Neumann–Morgenstern game theory for systems that are heavily interconnected. To see how this increase in complexity comes about, consider a system of N decision makers and suppose that the ith decision maker has n_i choices. A utility function for the ith decision maker thus has N independent variables, and the utility function make take on $n_1 n_2 \cdots n_N$ different values. Thus, a total of $N \prod_{i=1}^{N} n_i$ different values must be specified for the game. The interdependence function, however, has $2N$ independent variables, and therefore may assume $\prod_{i=1}^{N} n_i^2$ values. For example, suppose $N = 5$ and each decision maker can make only two choices, so $n_i = 2$. Each von Neumann–Morgenstern utility function may assume 32 values, generating a totality of 160 possible specifications for the entire system, whereas the interdependence function may require 1024 total specifications. This is the computational price that must be paid for accounting for preferences. Total preference interconnectivity does not come cheaply.

8.1   Game examples

This section illustrates, by means of examples, the increase in complexity that can occur with the adoption of a satisficing approach. Four well-known two-player games are discussed, and the satisficing approach is contrasted with the conventional von Neumann–Morgenstern approach. In each of these games, social relationships are modeled by parameters that may vary according to the emotional state of the players. The first game, "Bluffing," is a two-player zero-sum game of strict competition. The second game, "Battle of the Sexes," is a two-player coordination game in which the players' interests are compatible. The third game, "Prisoner's Dilemma," is a mixed-motive game involving notions of conflict and coordination. Finally, the fourth game, the "Ultimatum game," is a coordination game that exhibits apparently irrational behavior on the part of human players. These games have all been extensively analyzed from the point of view of von Neumann–Morgenstern game theory and thus make ideal examples of the difference between the classical approach and the satisficing approach.

8.1.1   Bluffing

Bluffing (von Neumann and Morgenstern, 1944; Bacharach, 1976) is a game involving two players, X_1 and X_2, who each stake themselves by putting on the table an ante,

Table 8.1: The payoff matrix
for the Bluffing game

	X_2	
X_1	F	C
S	0	$(b-a)/2$
B	a	0

a. X_1 then randomly draws one of two cards, marked Hi and Lo, from a hat, without revealing the result to X_2. X_1 has two choices. The first is to *fold* (F), in which case X_2 claims X_1's ante, and the game is over. Otherwise, X_1 must *raise* (R), that is, add an amount $b - a$ to the ante, making the total stake b. If X_1 raises, then X_2 must decide whether to fold, in which case X_1 claims X_2's ante, or X_2 must *call* (C), in which case X_1 must reveal the card that was drawn. X_1 wins b if the card is Hi, otherwise, X_2 wins b. X_1 has four possible strategies. Let them be denoted (F, F), (F, R), (R, F), and (R, R), where the first entry in the pair is X_1's response should the drawn card be Lo, and the second entry is the response should the drawn card be Hi. Clearly, (F, F) and (R, F) are dominated, and may be eliminated from consideration by a rational X_1. Notation may be simplified by relabeling the two non-dominated options for X_1 as playing straight, or $S = (F, R)$, and bluffing, or $B = (R, R)$. X_2's strategies are also simple: F or C. The payoff matrix for this game is given in Table 8.1, where each entry in the matrix corresponds to X_1's expected returns (rather than guaranteed returns).

The classical von Neumann–Morgenstern approach to this game is for each player to maximize its security level, that is, to maximize the minimum expected payoff. This principle leads to the notion of non-cooperative, or Nash, equilibria, that is, pairs of independently made choices such that neither player has motive to change, provided that the other player does not change. The minimax theorem establishes the fact that all finite two-player, zero-sum games have an equilibrium pair of pure or mixed strategies. For the bluffing game, the equilibrium is a mixed strategy: X_1 should play S with probability $\frac{2a}{a+b}$, and X_2 should play F also, it turns out, with probability $\frac{2a}{a+b}$. The minimax point, or value, of the game, is $\frac{b-a}{b+a}a$. This is the value that, on average, X_1 would win (and X_2 would lose) if the game were played many times. If a referee were to arbitrate, it would be fair simply to require X_2 to pay X_1 this amount in lieu of playing the game.

Although players would ostensibly participate in this game for the sole purpose of acquiring wealth and thus would presumably optimize, the fact that the game is not a fair one suggests that there may be ulterior motives. This is particularly relevant to X_2, since it has a higher likelihood of losing than winning, even under optimal play. Although any such ulterior motives are not provided by the expected utilities that appear in the payoff matrix, let us assume that the following social relationships exist: suppose that part of X_2's motive for playing is to embarrass its opponent by calling the other's bluff, and that X_1 mitigates its desire to win by its aversion to being dishonest, that is, to bluffing. We may capture these considerations by parameterizing the intensity of these relationships.

To cast this as a satisficing game, we must first establish operational definitions for failure avoidance and resource conservation. In accordance with the policy of associating the avoidance of failure with the fundamental goal of the game, this outcome corresponds to winning; any move that wins the game avoids failure. Resource conservation will be associated with taking risk (i.e., for X_1 to bluff and for X_2 to call). In accordance with the standard assumption associated with praxeic utility theory, we will assume that winning and taking risk are independent concepts for each individual; that is, X_1's bluffing risk and X_2's calling risk are independent, respectively, from each player's winning.

Our approach is to construct an interdependence function and to examine both joint and individual satisficing behavior. To compute the joint behavior, we invoke (6.12) and (6.13) and form a jointly satisficing set. We proceed by expressing the interdependence function as the following product (in the interest of brevity, we omit the arguments):

$$p_{S_1 S_2 R_1 R_2} = p_{R_1 | S_1 S_2 R_2} \cdot p_{R_2 | S_1 S_2} \cdot p_{S_2 | S_1} p_{S_1}. \tag{8.1}$$

Since failure avoidance and resource conservation are independent concepts for each player, it is immediate in this case that

$$p_{R_1 | S_1 S_2 R_2}(v_1 | u_1, u_2; v_2) = p_{R_1 | S_2 R_2}(v_1 | u_2; v_2)$$

and

$$p_{R_2 | S_1 S_2}(v_2 | u_1, u_2) = p_{R_2 | S_1}(v_2 | u_1).$$

We may simplify things further by observing that, conditioned on knowing X_2's preference for risk taking (calling), X_1 may calculate its preference for risk taking (bluffing) without considering its preference for winning. Accordingly, we have

$$p_{R_1 | S_2 R_2}(v_1 | u_2; v_2) = p_{R_1 | R_2}(v_1 | v_2),$$

and we may simplify (8.1) to become

$$p_{S_1 S_2 R_1 R_2}(u_1, u_2; v_1, v_2) = p_{R_1 | R_2}(v_1 | v_2) \cdot p_{R_2 | S_1}(v_2 | u_1) \cdot p_{S_2 | S_1}(u_2 | u_1) \cdot p_{S_1}(u_1). \tag{8.2}$$

To see how to generate $p_{R_1 | R_2}$, let us examine $p_{R_1 | R_2}(S | F)$, that is, the rejectability X_1 should associate with the option S, given that X_2 commits to rejecting F. In this situation, X_1 knows that X_2 will call, so it is not safe for X_1 to raise unless it draws Hi. Consequently, X_1 rejects B. Similarly, if X_1 knows that X_2 rejects C, X_1 knows it is safe to bluff and so will reject S. Summarizing, we have

$$
\begin{aligned}
p_{R_1|R_2}(S|F) &= 0, \\
p_{R_1|R_2}(B|F) &= 1, \\
p_{R_1|R_2}(S|C) &= 1, \\
p_{R_1|R_2}(B|C) &= 0.
\end{aligned}
\tag{8.3}
$$

Let us next examine the conditional selectability mass function $p_{R_2|S_1}(v_2|u_1)$. Suppose $u_1 = S$. Under this circumstance, X_2 knows that X_1 will not bluff, and so X_2 must reject calling, so $p_{R_2|S_1}(C|S) = 1$. Now suppose $u_1 = B$, that is, X_2 knows that X_1 will bluff. Let β denote the rejectability of folding in this case. Setting $\beta \approx 0$ means that X_2 ascribes low rejectability to folding and consequently a high rejectability to calling X_1's bluff. The resulting conditional rejectability is

$$
\begin{aligned}
p_{R_2|S_1}(F|S) &= 0, \\
p_{R_2|S_1}(C|S) &= 1, \\
p_{R_2|S_1}(F|B) &= \beta, \\
p_{R_2|S_1}(C|B) &= 1 - \beta.
\end{aligned}
\tag{8.4}
$$

Let us next consider the conditional selectability mass function $p_{S_2|S_1}(u_2|u_1)$. Suppose $u_1 = S$, that is, X_1 commits to playing straight. Knowing this, X_2 should take no risk, and must fold. If, however, $u_1 = B$, then X_2 knows that X_1 will raise, no matter what X_1 draws. Because of the symmetry in this game, it is obvious that X_2's attitude toward calling, knowing that X_1 will bluff, must be the same as X_2's attitude toward rejecting folding, knowing that X_1 will bluff. Thus, we have

$$
\begin{aligned}
p_{S_2|S_1}(F|S) &= 1, \\
p_{S_2|S_1}(C|S) &= 0, \\
p_{S_2|S_1}(F|B) &= 1 - \beta, \\
p_{S_2|S_1}(C|B) &= \beta.
\end{aligned}
\tag{8.5}
$$

Finally, we must construct X_1's myopic selectability, p_{S_1}. Let α denote X_1's selectability of bluffing.

$$
\begin{aligned}
p_{S_1}(S) &= 1 - \alpha, \\
p_{S_1}(B) &= \alpha.
\end{aligned}
\tag{8.6}
$$

The values of the interdependence function as parameterized by α and β are summarized in Table 8.2. We see that, whereas the von Neumann–Morgenstern approach requires the specification of eight parameters, which reduces to four parameters, as illustrated in Table 8.1, under the zero-sum assumption, the satisficing approach requires the specification of 16 values, albeit parameterized by only two quantities.

Table 8.2: The interdependence function for the Bluffing game

(u_1, u_2, v_1, v_2)	$p_{S_1 S_2 R_1 R_2}$	(u_1, u_2, v_1, v_2)	$p_{S_1 S_2 R_1 R_2}$
(S, F, S, F)	0	(B, F, S, F)	0
(S, F, S, C)	$1 - \alpha$	(B, F, S, C)	$\alpha(1 - \beta)^2$
(S, F, B, F)	0	(B, F, B, F)	$\alpha\beta(1 - \beta)$
(S, F, B, C)	0	(B, F, B, C)	0
(S, C, S, F)	0	(B, C, S, F)	0
(S, C, S, C)	0	(B, C, S, C)	$\alpha\beta(1 - \beta)$
(S, C, B, F)	0	(B, C, B, F)	$\alpha\beta^2$
(S, C, B, C)	0	(B, C, B, C)	0

Applying (6.12) and (6.13) yields

$$
\begin{aligned}
p_{S_1 S_2}(S, F) &= 1 - \alpha, \\
p_{S_1 S_2}(S, C) &= 0, \\
p_{S_1 S_2}(B, F) &= \alpha(1 - \beta), \\
p_{S_1 S_2}(B, C) &= \alpha\beta,
\end{aligned}
\tag{8.7}
$$

and

$$
\begin{aligned}
p_{R_1 R_2}(S, F) &= 0, \\
p_{R_1 R_2}(S, C) &= 1 - \alpha\beta, \\
p_{R_1 R_2}(B, F) &= \alpha\beta, \\
p_{R_1 R_2}(B, C) &= 0.
\end{aligned}
\tag{8.8}
$$

To determine the jointly satisficing set, we first must specify the boldness, q. For this game, it is reasonable that $q = 1$, thus ascribing equal weight to winning and taking risk. For $0 < \alpha < 1$ and $0 < \beta < 1$, the jointly satisficing set is

$$
\Sigma_q = \begin{cases}
\Sigma_q^1 = \{(S, F), (B, C)\} & \text{if } \beta > \tfrac{1}{2}, \\
\Sigma_q^2 = \{(S, F), (B, F), (B, C)\} & \text{if } \beta \leq \tfrac{1}{2}.
\end{cases}
$$

This set may be viewed as a list of possibilities that, if followed by both players, would generate results that are jointly "good enough" for both of them, where "good enough" means that the positive attributes of the action (the joint support for winning) equals or exceeds the negative attributes (the risk incurred by bluffing and calling).

Bluffing is a game of pure conflict, and it is not expected that players would wish to adopt a joint decision. Thus, to complete our analysis we must compute the individually satisficing solutions. The individual selectability function for X_1 is computed as

$$
p_{S_1}(u) = \overset{\cdot}{p}_{S_1 S_2}(u, F) + p_{S_1 S_2}(u, C)
$$

for $u \in \{S, B\}$, with a similar calculation required to obtain X_1's individual rejectability, and X_2's individual selectability and rejectability. The resulting functions are

$$p_{S_1}(S) = 1 - \alpha, \quad p_{R_1}(S) = 1 - \alpha\beta,$$
$$p_{S_1}(B) = \alpha, \qquad p_{R_1}(B) = \alpha\beta,$$

and

$$p_{S_2}(F) = 1 - \alpha\beta, \quad p_{R_2}(F) = \alpha\beta,$$
$$p_{S_2}(C) = \alpha\beta, \qquad p_{R_2}(C) = 1 - \alpha\beta.$$

The resulting univariate satisficing sets are

$$\Sigma_q^1 = \begin{cases} \{B\} & \text{for } \beta < 1, \\ \{S, B\} & \text{for } \beta = 1, \end{cases}$$

$$\Sigma_q^2 = \begin{cases} \{F\} & \text{for } \alpha\beta < \frac{1}{2}, \\ \{C\} & \text{for } \alpha\beta > \frac{1}{2}, \\ \{F, C\} & \text{for } \alpha\beta = \frac{1}{2}, \end{cases}$$

and the satisficing rectangle is

$$\mathfrak{R}_q = \Sigma_q^1 \times \Sigma_q^2 = \begin{cases} \{B, F\} & \text{for } \alpha\beta < \frac{1}{2}, \beta < 1, \\ \{B, C\} & \text{for } \alpha\beta > \frac{1}{2}, \beta < 1, \\ \{\{B, F\}, \{B, C\}\} & \text{for } \alpha\beta = \frac{1}{2}, \beta < 1, \\ \{\{B, F\}, \{B, C\}, \{S, F\}, \{S, C\}\} & \text{for } \alpha\beta = \frac{1}{2}, \beta = 1, \\ \{\{B, F\}, \{S, F\}\} & \text{for } \alpha\beta < \frac{1}{2}, \beta = 1, \\ \{\{B, C\}, \{S, C\}\} & \text{for } \alpha\beta > \frac{1}{2}, \beta = 1. \end{cases}$$

There is a marked difference between the satisficing solution and the traditional minimax solution. With the satisficing approach, both players are content with breaking even; whereas, the minimax approach presents a clear advantage to X_1. But the minimax approach is not without its problems. Bacharach (1976) identifies a desirable property of any choice: *if both players use the same choice principle, neither will afterwards regret having used it.* If the game is to be played repeatedly, we may interpret the probabilities in a mixed strategy as long-run relative frequencies, and randomizing according to the optimal distribution is a way to prevent an intelligent opponent from learning one's intentions. It seems plausible, under these circumstances, that the minimax expected utility principle is rational. For one-off games (games played once), however, long-run relative frequencies are not relevant, and one must adopt a different justification for playing a mixed strategy. If a player employs the minimax expected utility principle for a one-off game, it *is* possible to regret doing so (for example, X_1 bluffs when drawing *Lo*, and X_2 calls). Thus, it seems that the maximin strategy is problematical as an optimal strategy for one-off scenarios.

Even for one-off games, why would X_1 ever settle for a satisficing solution when it has at least a probabilistic advantage by choosing a random strategy according to the optimal distribution defined by the minimax theorem? I do not have an easy answer. On the one hand, the optimizer bases its decision on the expected returns without consideration of any "personality" factors, such as the players' propensities regarding bluffing and folding. Indeed, any such propensities are dictated to the players by the optimal distributions – the decision makers are assumed to be completely dispassionate. On the other hand, the satisficing decision criteria are structured at least partly in terms of preferences that are related to the players' "personalities," namely, their respective propensities to bluff and fold. It should be noted, however, that the interdependence function itself is structured in accordance with the goals of the players to win the game (as the payoff matrix under the conventional formulation); only the parameters α and β are left as free variables to be specified according to the propensities of the players.

Another question is: why would a rational X_2 ever consent to play the game, know-ing that it is more likely to lose than to win?[2] Game theory always seems to as-sume that the game will be played, even if under duress. While this may essentially be true in situations where games are used as models of actual economic behavior, compulsion is not a factor in the literal application of the theory, such as the card game under discussion. An optimizing X_2 should never willingly play the game in the first place.

8.1.2   Battle of the Sexes

As introduced in Chapter 1 and repeated here in the interest of continuity, Battle of the Sexes[3] is a game involving a man and a woman who plan to meet in town for a social function. She (S) prefers to go to the ballet (B), while he (H) prefers the dog races (D). Each prefers to be with the other, however, wherever the social function may be. The payoff matrix for this game is given in Table 1.2 in ordinal form.

An approach that does not require communication is for each player to flip a coin and choose according to the outcome of that randomizing experiment. If they were to repeat this game many times, then, on average, each player would realize an outcome midway between next best and next worst for each. But, for any given trial, they would be in one of the four states with equal probability.

If the players could communicate, then a much better strategy would be to alternate between (D, D) and (B, B), thus ensuring an average level of satisfaction midway between best and next best for each. If they possess the ability to learn from experience, they may also converge to an alternation scheme under repeated play.

---

[2] Of course, casinos depend upon such behavior, but who would argue that betting against the house is rational?
[3] This game was widely discussed long before sexist issues became sensitive, and cell-phones became available. I hope readers will not be offended by the stereotypical roles assumed by the players.

Table 8.3: The payoff matrix for the
Distributed Manufacturing game

	X_2	
X_1	Shades	Cloths
Lamps	($20, $5)	($10, $4)
Tables	($8, $3)	($15, $10)

Regardless of the strategies that may be employed, this game, as it is configured by the payoff matrix, illustrates the shortcomings of conventional utility theory for the characterization of behavior when cooperation is essential. Each player's level of satisfaction is determined completely as a function if his or her own enjoyment. For example, the strategy vector (D, D) is best for H, but it is because he gets his way on both counts: he goes to his favorite event and he is with S. Her feelings, however, are not taken into consideration. According to the setup, it would not matter to H if S were to detest dog races and were willing to put up with that event at great sacrifice of her own enjoyment, just to be with H. Such selfish attitudes, though not explicit, are at least implied by the structure of the payoff matrix, and are likely to send any budding romance to the dogs. The problem is that the solution concept, based as it is upon individual rationality, fosters competition, even though cooperation is desired.

As an illustration of a subtle type of competition that may emerge from this game, consider a distributed manufacturing game involving the the operation of a shop floor. Producer X_1 can manufacture lamps or tables, and Producer X_2 can manufacture lamp shades or table cloths, but each must choose which product to manufacture without direct communication. Coordinated behavior would make both of their products more marketable, as indicated in Table 8.3, which displays the net profit accruing to each producer as a function of their joint decisions. Clearly, this is an instantiation of the Battle of the Sexes game.

Using these numbers, X_1 might reason that, since his profit for (Lamps, Shades) is twice the profit to X_2 for (Tables, Cloths) but the incremental change in profit is the same for both, then his preference is stronger and should prevail. On the other hand, however, X_2 might reason that, since it is worth twice as much to her if they produce (Tables, Cloths), rather than (Lamps, Shades) and it is only $\frac{4}{3}$ more valuable to X_1 if they produce (Lamps, Shades) rather than (Tables, Cloths), her preference is stronger and should prevail. Of course, if side-payments are allowed, then this could help resolve the dilemma, but that would also create a secondary game of how to arrive at equitable side-payments – which might produce an infinite regress of dilemmas.

Let us cast the Battle of the Sexes as a satisficing game. We must first establish each player's notions of failure avoidance and resource conservation. In accordance with our previous discussion, the former is related to the most important goal of the game, which is for the two players to be with each other, regardless of where they go. Resource conservation, on the other hand, deals with the costs of being at a particular function. Obviously, H would prefer D if he did not take into consideration S's preferences; similarly, S would prefer B. Thus, we may express the myopic rejectabilities for H and S in terms of parameters h and s, respectively, as

$$
\begin{aligned}
p_{R_H}(D) &= h, \\
p_{R_H}(B) &= 1 - h,
\end{aligned}
\tag{8.9}
$$

and

$$
\begin{aligned}
p_{R_S}(D) &= 1 - s, \\
p_{R_S}(B) &= s,
\end{aligned}
\tag{8.10}
$$

where h is H's rejectability of D and s is S's rejectability of B. The closer h is to zero, the more H is adverse to B with an analogous interpretation for s with respect to S attending D. To be consistent with the stereotypical roles, we may assume that $0 \leq h < \frac{1}{2}$ and $0 \leq s < \frac{1}{2}$. As will be subsequently seen, only the ordinal relationship need be specified, that is, either $s < h$ or $h < s$.

Selectability is a measure of the failure avoidance associated with the options. Since being together is a joint, rather than an individual objective, it is difficult to form unilateral assessments of selectability, but it is possible to characterize individually the conditional selectability. To do so requires the specification of the conditional mass functions $p_{S_H | R_S}$ and $p_{S_S | R_H}$; that is, H's selectability conditioned on S's rejectability and S's selectability conditioned on H's rejectability. If S were to place her entire unit mass of rejectability on D, H might account for this, if he cares at all about S's feelings, by placing some portion of his conditional selectability mass on B. S might construct her conditional selectability in a similar way, yielding

$$
\begin{aligned}
p_{S_H | R_S}(D|D) &= 1 - \alpha, \\
p_{S_H | R_S}(B|D) &= \alpha, \\
p_{S_H | R_S}(D|B) &= 1, \\
p_{S_H | R_S}(B|B) &= 0,
\end{aligned}
\tag{8.11}
$$

and

$$
\begin{aligned}
p_{S_S | R_H}(D|D) &= 0, \\
p_{S_S | R_H}(B|D) &= 1, \\
p_{S_S | R_H}(D|B) &= \beta, \\
p_{S_S | R_H}(B|B) &= 1 - \beta.
\end{aligned}
\tag{8.12}
$$

The valuations $p_{S_H | R_S}(B|D) = \alpha$ and $p_{S_S | R_H}(D|B) = \beta$ can be considered conditions of situational altruism. If S were to place all of her rejectability mass on D, then H might defer to S's strong dislike of D by placing α of his selectability mass, *as conditioned by her preference*, on B. Similarly, S could show a symmetric conditional preference for D if H were to reject B strongly. The parameters α and β are H's and S's **indices of situational altruism**, respectively, and serve as a way for each to control the amount of deference he or she is willing to grant to the other. In the interest of simplicity, we shall assume that both players are maximally altruistic and set $\alpha = \beta = 1$. In principle, however, they may be set independently to any value in $[0, 1]$. Notice that, even in this most altruistic case, these conditional preferences do not commit one to categorical abdication of his or her own unilateral preferences. H still myopically (that is, without taking S into consideration) prefers D, and S still myopically prefers B, and there is no intimation that either participant must throw the game in order to accommodate the other.

With these conditional and marginal functions, we may factor the interdependence function as follows:

$$p_{S_H S_S R_H R_S}(x, y; z, w) = p_{S_H | S_S R_H R_S}(x | y; z, w)$$
$$\cdot p_{S_S | R_H R_S}(y | z, w) \cdot p_{R_H R_S}(z, w)$$
$$= p_{S_H | R_S}(x | w) \cdot p_{S_S | R_H}(y | z)$$
$$\cdot p_{R_H}(z) \cdot p_{R_S}(w),$$

where we have assumed that H's selectability conditioned on S's rejectability is dependent only on S's rejectability, that S's selectability conditioned on H's rejectability is dependent only on H's rejectability, and that the myopic rejectability values of H and S are independent.

Application of (6.12) and (6.13) results in joint selectability and rejectability values of

$$p_{S_H S_S}(D, D) = (1 - h)s,$$
$$p_{S_H S_S}(D, B) = hs,$$
$$p_{S_H S_S}(B, D) = (1 - h)(1 - s),$$
$$p_{S_H S_S}(B, B) = h(1 - s),$$

$$(8.13)$$

and

$$p_{R_H R_S}(D, D) = h(1 - s),$$
$$p_{R_H R_S}(D, B) = hs,$$
$$p_{R_H R_S}(B, D) = (1 - h)(1 - s),$$
$$p_{R_H R_S}(B, B) = (1 - h)s.$$

$$(8.14)$$

The marginal selectability and rejectability values for H and S are

$$p_{S_H}(D) = s, \qquad p_{R_H}(D) = h, \tag{8.15}$$
$$p_{S_H}(B) = 1 - s, \qquad p_{R_H}(B) = 1 - h, \tag{8.16}$$

and

$$p_{S_S}(D) = 1 - h, \qquad p_{R_S}(D) = 1 - s, \tag{8.17}$$
$$p_{S_S}(B) = h, \qquad p_{R_S}(B) = s. \tag{8.18}$$

Setting the index of caution, q, equal to unity, we obtain the jointly satisficing set as

$$\Sigma_q = \begin{cases} \{(D, B), (B, D), (B, B)\} & \text{for } s < h, \\ \{(D, D), (D, B), (B, D)\} & \text{for } s > h, \\ \{(D, D), (D, B), (B, D), (B, B)\} & \text{for } s = h, \end{cases}$$

the individually satisficing sets are

$$\Sigma_q^H = \begin{cases} \{B\} & \text{for } s < h, \\ \{D\} & \text{for } s > h, \\ \{B, D\} & \text{for } s = h, \end{cases}$$

$$\Sigma_q^S = \begin{cases} \{B\} & \text{for } s < h, \\ \{D\} & \text{for } s > h, \\ \{B, D\} & \text{for } s = h, \end{cases}$$

and the satisficing rectangle is

$$\Re_q = \Sigma_q^H \times \Sigma_q^S = \begin{cases} \{B, B\} & \text{for } s < h, \\ \{D, D\} & \text{for } s > h, \\ \{\{B, B\}, \{D, D\}\} & \text{for } s = h. \end{cases}$$

Thus, if S's aversion to D is less than H's aversion to B, then both players will go to H's preference, namely, D, and conversely. This interpretation is an example of interpersonal comparisons of utility. As discussed in Chapter 5, such comparisons are frowned upon by conventional game theorists, but are essential to social choice theory, so long as the utilities are expressed in the same units and have the same zero-level. Since the utilities are mass functions, however, the analysis in Section 5.3 applies. We observe, for example, that if $s < h$ then H is ambivalent with respect to B and S is dubious with respect to D. Consequently, according to the analysis of Section 5.4, these preferences are not invariant under isomassive transformations, and it would be possible to reorder the preferences by such transformations. Thus, it is imperative that both players have reliable assessments of the strength of their own and their partner's preferences relative to each other.

It is useful to discuss structural differences between the representation of this game as a traditional game and its representation as a satisficing game. With the traditional game, all of the information is encoded in the payoff matrix, with the values that are assigned representing the importance, cost, or informational value to the players. The payoff matrix characterizes the gains or losses to a player conditioned on the action of itself and the other player. For example, suppose values were assigned to the ordinal payoff matrix given in Table 1.2, resulting in a numerical payoff matrix given in Table 8.4, where the values are in cardinal units of, say, "satisfaction." We would be justified in assuming that S would be twice as satisfied, given that she is with H, to be at B rather than at D. Also, given that she would gain one unit of satisfaction in that

Table 8.4: A numerical payoff matrix for the Battle of the Sexes game

		S
H	D	B
D	(2, 1)	(−1, −1)
B	(−2, −2)	(1, 2)

way, she would experience that same amount of dissatisfaction were she alone at B, and twice that amount if she were alone at D.

With the satisficing game structure, all of the information is encoded into the interdependence function. This function may be factored into products of conditional interdependencies that represent the joint and individual goals and preferences of the players. The joint selectability function (8.13) and the joint rejectability function (8.14) characterize the state of the problem as represented by the conditional goals and individual preferences of the the players. Specifying numerical values for the preference parameters, h and s, is as natural, it may be argued, as it would be to specify numerical values for the payoff matrix.

The essential difference between the von Neumann–Morgenstern and the satisficing representations of this game is that the von Neumann–Morgenstern utilities do not permit the preferences of one player to influence the preferences of the other player. *But in the context of the game, it is reasonable to assume that such preferences exist. If they do, then there should be a mechanism to account for them. Classical game theory does not provide such a mechanism, but satisficing game theory does, at the expense of added complexity.*

Another important way to compare these two game formulations is in terms of the solution concept. The classical solution to the traditional problem is to solve for mixed strategy Nash equilibria corresponding to a numerically definite payoff matrix. The main problem with using a mixed strategy, as far as this discussion is concerned, is that, for it to be successful, both players must use exactly the same payoff matrix, and they must use the randomized decision probabilities that are calculated to maximize their payoffs. Even small deviations in these quantities destroy the equilibrium completely and the solution has no claim on rationality, let alone optimality. The reason for this sensitivity is that the players are attempting to optimize and, to do so, they must exploit the model structure and parameter values to the maximum extent possible.

The satisficing solution concept, on the other hand, adopts a completely different approach to decision making. There is no explicit attempt to optimize. Rather than ranking all of the possible options according to their expected utility, the attributes (selectability and rejectability) of each option are compared and a binary decision is made with respect to each option, and it is either rejected or it is not. For this problem,

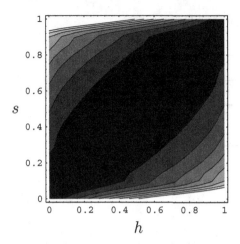

Figure 8.1: The contour plot of the diversity functional for the Battle of the Sexes game.

the comparison was made in terms of the parameters s and h, and all that is important is their ordinal relationship; numerically precise values would not necessarily be exploited were they available.

It is instructive to evaluate the joint diversity function for this problem as a function of the parameters h and s. From (8.13) and (8.14), we may express the diversity function as

$$D(h\|s) = (1 - h)s \log \frac{(1 - h)s}{h(1 - s)} + h(1 - s) \log \frac{h(1 - s)}{(1 - h)s}.$$

Figure 8.1 illustrates this quantity. Note, in keeping with intuition, that the fitness significantly improves as the difference between h and s increases.

This game illustrates the common-sense attribute of the satisficing solution without the need for randomizing decisions or for S to guess what H is guessing S is guessing, and so forth, *ad infinitum*. The players go to the function that is considered by both to be the less rejectable. This decision is easily justified by common-sense reasoning. Furthermore, the satisficing solution is more robust than than the conventional solution. There is no need to define numerically precise payoffs, and there is no need to calculate precise probability distributions according to which random decisions will be chosen. Finally, there is no need to specify more than an ordinal relationship regarding player preferences to arrive at a decision.

To apply this result to the Distributed Manufacturing game illustrated in Table 8.3, we interpret h as X_1's rejectability of producing lamps and s as X_2's rejectability of producing table cloths. One reasonable way to compute these rejectabilities is to argue that the rejectability ratio for each player of its two options ought to be the reciprocal of the individual profit ratios for the two cooperative solutions. This approach yields the ratios $\frac{h}{1-h} = \frac{15}{20}$ and $\frac{s}{1-s} = \frac{5}{10}$, or $h = \frac{3}{7}$ and $s = \frac{1}{3}$. Since $s < h$ in this case, the only jointly and individually satisficing solution is (Tables, Cloths).

Table 8.5: The payoff matrix in ordinal form for the Prisoner's Dilemma game

		X_2
X_1	C	D
C	(3, 3)	(1, 4)
D	(4, 1)	(2, 2)

Key: 4 = best; 3 = next best; 2 = next worst; 1 = worst

8.1.3   Prisoner's Dilemma

One of the most famous of all games is the Prisoner's Dilemma. This game involves two players, X_1 and X_2, who have been charged with a serious crime, arrested, and incarcerated in a way that precludes any communication between them. The prosecution has evidence sufficient only to convict them of a lesser crime with a moderate jail sentence. To get at least one conviction on the more serious crime, the prosecution entices each prisoner to give evidence against the other, that is, to defect (D). Otherwise, each prisoner may cooperate (C) with the other by not supplying evidence. Defection yields dropped charges if the other prisoner cooperates; cooperation yields the maximum sentence if the other defects. If both cooperate, both receive short sentences; if both defect, both receive moderate sentences.

The classical von Neumann–Morgenstern solution is obtained in terms of a payoff matrix, as illustrated in ordinal form in Table 8.5. Clearly, playing D for either player is the dominant strategy. Furthermore, (D, D) is also the unique Nash equilibrium pair. Unfortunately, this solution is inferior to playing the Pareto equilibrium solution, (C, C), as it results in the next-worst, rather than the next-best, consequence.

One of the characteristics of the von Neumann–Morgenstern approach is that it abstracts the game from its context, or story line – all relevant information is captured by the utility functions. These utilities represent individual preferences as functions of joint actions; they do not represent joint preferences. It is only when the two utility functions are juxtaposed in the payoff matrix that the "game" emerges and strategies can be devised. Under this view, the players are assumed to be absolutely certain that self-interest is the only issue. The von Neumann–Morgenstern approach does not countenance mixed motives. Accounting for any dispositions for coordinated behavior would change the payoff matrix and, hence, the game.[4] There is no room for equivocation.

---

[4] It is interesting to note that the story that gives the Prisoner's Dilemma its name was invented *post hoc* to conform, for pedagogical illustration, to the payoff matrix with the given structure, rather than the other way around (Straffin, 1980). This illustrates the mind-set of many game theorists: the actual "game" is the payoff matrix, not the story line.

This is a powerful, but necessary, assumption under the von Neumann–Morgenstern approach to this game.

It is interesting to examine repeated-play versions of this game from the heuristic point of view. Of course, the standard von Neumann–Morgenstern approach must return the Nash solution, no matter how many times the game is played, but other approaches have revealed quite different behavior. An interesting approach was introduced by Rapoport, who proposed a simple *tit-for-tat* rule of repeated play: start by cooperating, thereafter play what the other player chose in the previous round. This purely heuristic rule won the Axelrod Tournament (Axelrod, 1984). One of its main characteristics is that it does not abstract the game from the story – the actions are taken completely within the game context.

The von Neumann–Morgenstern and Rapoport approaches to this game represent two extremes. With the von Neumann–Morgenstern approach, all relevant knowledge is encoded into the expected utility functions, self-interest is the only issue, and expected utility maximization is the operative solution concept. With the Rapoport approach, the players' dispositions enter into the decision, the solution concept is an *ad hoc* rule to be followed, and there are no attempts to rank-order options or maximize performance.[5]

Let us apply satisficing game theory to the Prisoner's Dilemma game. Our task is to define the interdependence mass function, from which the selectability and rejectability mass functions can be obtained and compared for each joint option. To generate the interdependence mass function, we must first define operational notions for failure avoidance and resource conservation.

The fundamental objective of the players, as expressed by their utility functions, is to get out of jail quickly – self-interest is the primary consideration. Laboratory experiments with randomly selected humans, however, result in the cooperative solution being chosen relatively frequently with single play and no communication between the participants (Wolf and Shubik, 1974). This evidence suggests that, for the game to be a model of human behavior, motives may exist in addition to self-interest. The individual utilities displayed in Table 8.5, however, do not accommodate any other interests, and there is no way to build anything but self-interest into the conventional formulation of the game. The satisficing approach, however, requires the formulation of two utility functions, thus opening up the possibility of accommodating issues in addition to exclusive self-interest.

One possibility is group interest, which apparently emerges in repeated play as a result of learned cooperation.[6] We view the players of this game as individuals who are

---

[5] Traditional game theorists point out that Rapoport's solution is not optimal, but Axelrod (1984) has shown that, for repeated-play games where future payoffs are important, there does not exist an optimal strategy that is independent of the strategies used by other players.

[6] Although other psychological factors, such as the other's expected behavior, may contribute to the results of repeated-play games, for simplicity we restrict attention to the notion of group interest.

concerned primarily with their own task, but at the same time have a degree of consideration for other players' difficulties and consider it a cost to them if an action they take makes it difficult for others. Such players are enlightened liberals, as described in Section 7.2. Accordingly, we proceed by associating failure avoidance with reducing individual jail time, and associating resource conservation with group jail time. Thus, short individual sentences will have high individual selectability, and long group sentences will have high joint rejectability.

The notion of group interest can have significance only if the players each acknowledge some form of dependence on the other, however weak it may be. To the extent that these dependencies reinforce each other, the players implicitly forge a joint opinion regarding the relative merits of cooperation and conflict. The success of Rapoport's approach suggests that there may be (at least) two attitudes in the minds of the players that may affect their decisions: (a) a propensity for *dissociation*, that is, for the players to go their separate ways without regard for coordination, and (b) a propensity for *vulnerability*, that is, for the players to expose themselves to individual risk in the hope of improving the joint outcome.

Let $\alpha \in [0, 1]$ be a measure of the joint value the players place on rejecting the joint option (C, C). We may identify α as a *dissociation index:* if $\alpha \approx 1$, the players are completely dissociated and coordination is unlikely. Also, let $\beta \in [0, 1]$ be a measure of the joint value placed on rejecting the joint option (D, D). β may be viewed as a *vulnerability index:* $\beta \approx 1$ means the players are each willing to risk a long jail sentence in the hope of both obtaining a shorter one. A condition of high dissociation *and* high vulnerability would indicate lack of concern for the welfare of the other player while at the same time implying a willingness to expose oneself to dire consequences. We may prohibit this situation by imposing the constraint that $\alpha + \beta \leq 1$. If, for example, $\alpha = 1$ and $\beta = 0$, then self-interest is the only consideration. If, however, $\alpha = 0$ and $\beta = 1$, then the players are willing to assume high risk to achieve cooperation.

If the sentence lengths are independent of the identity of the players, then the joint rejectability of (D, C) should be equal to the joint rejectability of (C, D). With this assumption and the constraints on α and β, we define the joint rejectability mass function:

$$p_{R_1 R_2}(C, C) = \alpha, \qquad p_{R_1 R_2}(C, D) = \frac{1 - \alpha - \beta}{2},$$

$$p_{R_1 R_2}(D, C) = \frac{1 - \alpha - \beta}{2}, \qquad p_{R_1 R_2}(D, D) = \beta. \tag{8.19}$$

Although selectability deals with individual objectives, it is a joint consideration, since the consequences of the players' decisions are not independent, and thus preferences cannot be independent. A convenient way to express this dependency is to exploit the probabilistic structure of the interdependence mass function and to compute joint

Table 8.6: The conditional selectability $p_{S_1 S_2 | R_1 R_2}(v_1, v_2 | w_1, w_2)$ for the Prisoner's Dilemma game

	(w_1, w_2)			
(v_1, v_2)	(C, C)	(C, D)	(D, C)	(D, D)
(C, C)	0	0	0	1
(C, D)	0	0	0	0
(D, C)	0	0	0	0
(D, D)	1	1	1	0

selectability conditioned on joint rejectability,

$$p_{S_1 S_2 | R_1 R_2}(v_1, v_2 | w_1, w_2)$$

for all $(v_1, v_2) \in \mathbf{U}$ and all $(w_1, w_2) \in \mathbf{U}$, where

$$\mathbf{U} = U \times U = \{(C, C), (C, D), (D, C), (D, D)\}.$$

The interdependence mass function may then be obtained by the product rule:

$$p_{S_1 S_2 R_1 R_2}(v_1, v_2; w_1, w_2) = p_{S_1 S_2 | R_1 R_2}(v_1, v_2 | w_1, w_2) \cdot p_{R_1 R_2}(w_1, w_2). \tag{8.20}$$

The function $p_{S_1 S_2 | R_1 R_2}(v_1, v_2 | w_1, w_2)$ characterizes the selectability of the joint option (v_1, v_2) given that the players jointly place all of their rejectability mass on (w_1, w_2). We may compute the conditional selectability by invoking straightforward and intuitive rules of the form: "If X_1 and X_2 jointly reject (w_1, w_2), then they should jointly select (v_1, v_2)." Let, say, $(w_1, w_2) = (C, C)$, that is, the players jointly reject cooperation. Given this situation, it is trivially obvious that the preferred joint option is for both to defect.[7] We may encode this rule into the conditional selectability mass function by placing all of the selectability mass on the joint option (D, D), that is, $p_{S_1 S_2 | R_1 R_2}(D, D | C, C) = 1$. If exactly one player rejects cooperating, then it is obvious that, in this case as well, the preferred joint option is for both to defect, consequently, we set $p_{S_1 S_2 | R_1 R_2}(D, D | C, D) = 1$ and $p_{S_1 S_2 | R_1 R_2}(D, D | D, C) = 1$. We complete this development by noting that, if both players reject defecting, then $p_{S_1 S_2 | R_1 R_2}(C, C | D, D) = 1$. Table 8.6 summarizes the structure of this conditional selectability function.

Substituting the conditional selectability function given by Table 8.6 and the joint rejectability given by (8.19) and (8.20) and applying (6.12) yields the joint selectability function:

$$\begin{aligned} p_{S_1 S_2}(C, C) &= \beta, & p_{S_1 S_2}(C, D) &= 0, \\ p_{S_1 S_2}(D, C) &= 0, & p_{S_1 S_2}(D, D) &= 1 - \beta. \end{aligned} \tag{8.21}$$

---

[7] This apparent triviality is a consequence of each player having only two options, since rejecting one implies selecting the other, but the situation is not so trivial when there are more than two options.

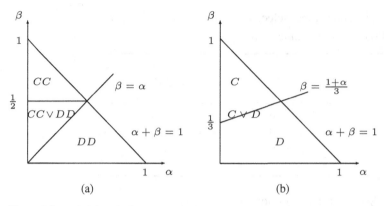

Figure 8.2: Satisficing decision regions for the Prisoner's Dilemma game: (a) bilateral decisions and (b) unilateral decisions.

Comparing the bilateral selectability, (8.21), with the bilateral rejectability, (8.19), we obtain the bilateral satisficing set, which consists of all decision pairs, $(w_1, w_2) \in \mathbf{U}$, such that

$$p_{S_1 S_2}(w_1, w_2) \geq q p_{R_1 R_2}(w_1, w_2).$$

Thus, parameterized by α and β, this set is, for the special case $q = 1$,

$$\Sigma_q = \begin{cases} \{(C, C)\} & \text{for } \beta \geq \frac{1}{2}, \\ \{(D, D)\} & \text{for } \beta \leq \alpha, \\ \{(C, C), (D, D)\} & \text{for } \alpha < \beta < \frac{1}{2}. \end{cases} \tag{8.22}$$

These regions are depicted in Figure 8.2(a). The bilateral satisficing set coincides with the Pareto equilibrium solution, (C, C), when the vulnerability index is at least as large as $\frac{1}{2}$. It coincides with the Nash solution when the vulnerability index is less than the dissociation index. If the vulnerability index is greater than the dissociation index but less than $\frac{1}{2}$, then the bilateral satisficing set contains both (C, C) and (D, D). To take action in this situation requires the invocation of a tie-breaker. For example, (D, D) is the satisficing option placing higher emphasis on individual interest (higher selectability), and (C, C) is the satisficing option placing higher emphasis on group interest (lower rejectability).

We may compute the unilateral satisficing set by computing the univariate selectability and rejectability marginals in accordance with (6.17) and (6.18), from which the univariate satisficing set for either player is

$$\Sigma_q = \begin{cases} \{C\} & \text{for } \beta > \frac{1+\alpha}{3}, \\ \{D\} & \text{for } \beta < \frac{1+\alpha}{3}, \\ \{C, D\} & \text{for } \beta = \frac{1+\alpha}{3}. \end{cases}$$

The unilateral decision regions are illustrated in Figure 8.2(b). Note that the set is a singleton except in the special situation of $\beta = \frac{1+\alpha}{3}$.

In contrast to the von Neumann–Morgenstern solution to this game, the satisficing solution accounts for the social inclinations of the players. The von Neumann–Morgenstern solution emerges as a special case (e.g., $\alpha = 1$ and $\beta = 0$), but the satisficing solution gives the solution for all admissible (α, β) pairs. Uncertainty regarding these parameters may be handled in several ways. A decision maker is free to (a) regard them as random variables with known or interval-valued distributions and compute expectations, (b) regard them as deterministic interval-valued parameters and perform worst-case analysis, or (c) regard them as unknown parameters for which only an ordinal relationship is assumed. Thus, under fairly general circumstances, a decision can be rendered even though there may be considerable uncertainty regarding the values of the social indices.

8.1.4   The Ultimatum game

The Ultimatum game is as follows: one player, called the proposer, offers another player, called the responder, a fraction of a fortune, and the responder chooses whether or not to accept the offer. If the responder accepts, then the two players divide the fortune between themselves according to the agreed-upon ratio, but if the responder declines the offer, neither player receives anything. In both cases, the game is over; there is no opportunity for reconsideration.

The von Neumann–Morgenstern solution to this game is for the proposer to offer the smallest non-zero amount it can, and for the responder to accept whatever is offered. This is the play predicted by individual rationality. Interestingly, such a strategy is rarely adopted by human players. Even with one-off play, proposers are inclined to offer fair deals, and responders are inclined to reject unfair deals (Roth, 1995).

There have been many attempts to explain this phenomenon, with the most popular ones being: (a) humans find it difficult to permit a rival to gain an advantage; (b) group dependency motivates players to maintain socially acceptable reputations; (c) people initially do no not fully understand the game and must learn to optimize through trial-and-error; and (d) players are more influenced by social norms than by strategic considerations such as optimization (Nowak et al., 2000; Sigmund et al., 2002; Binmore, 1998).

With all of these explanations except the last one, which has a distinctive heuristic flavor, the players have modified their utilities to account for social considerations, but individual rationality has not been replaced. They are still playing to maximize their expected utility; their utilities just happen to include issues in addition to financial payoffs. In other words, relevant social relationships exist but, since von Neumann–Morgenstern game theory is based solely on individual rationality, it does not readily accommodate them.

Two issues are relevant here. First, if social relationships are material considerations for the behavior of the players, it seems reasonable that such attributes should be

Table 8.7: The payoff matrix for the Ultimatum minigame

	X_2	
X_1	a	d
h	$(1-h, h)$	$(0, 0)$
ℓ	$(1-\ell, \ell)$	$(0, 0)$

explicitly included in the structure of the game, rather than being merely appended as *post factum* explanations for apparently anomalous behavior. Perhaps too much has been abstracted from the story line in the von Neumann–Morgenstern formulation of the game to ensure that a payoff array captures all of the essential attributes of the players.

The second issue is that, taking the results of the Ultimatum game at face value, it is difficult to cling to the hypothesis that people are always expected utility maximizers.[8] There is much empirical evidence to the contrary, and it is at least an open question. Consequently, reasonable alternative hypotheses deserve to be investigated.

The Ultimatum game permits the proposer to offer any fraction of the fortune to the responder, and thus the proposer has a continuum of options at her disposal, while the responder has only two options: a (accept), or d (decline). The game loses little of its effect, and its analysis is much simpler, if we follow the lead of Gale et al. (1995) and consider the so-called minigame, with only two possible offers, h and ℓ (high and low), with $0 < \ell < h \leq \frac{1}{2}$. These values correspond to the fraction of the fortune that the proposer, denoted X_1, is prepared to offer to the responder, denoted X_2. This minigame analysis captures the essential features of the continuum game and permits us to see clearly the relationships between the two players. With this restriction, the option sets for X_1 and X_2 are $U_1 = \{h, \ell\}$ and $U_2 = \{a, d\}$, respectively. The von Neumann–Morgenstern payoff matrix for the Ultimatum minigame is given in Table 8.7.

The unique Nash equilibrium for this game is for X_1 to play ℓ and X_2 to play a. According to the doctrine of individual rationality, the players should adopt this joint strategy. The response of many human players of this game, however, is an indication that there is more to this game than meets the eye, since players are more prone to play fair than to optimize. Evidently, greed on the part of proposers is often mitigated, causing them to make reasonably fair offers, and responders often refuse to accept an offer if it is too low, presumably because of envy, humiliation, or other emotional attributes. But if social considerations are important, they should be part of the game structure. There are (at least) two emotional aspects that appear to be relevant. First, if the responder's indignation is relevant to the game, a parameter representing this social attribute should be part of the mathematical structure of the game.

---

[8] The tenuous argument that people always maximize, but with respect to imprecisely known utilities, can neither be proven nor disproven.

A second potential social attribute is that the proposer may be motivated by considerations other than temerarious greed; namely, the pragmatic notion that the responder may reject the offer if it is extremely unfair, thus denying the proposer any benefit. Consequently, even if the proposer is avaricious, she may make an equitable offer if she suspects that the responder would be prone to reject an inequitable one, but if she suspected that the proposer were not so prone, she would make a inequitable offer. In other words, the proposer may be willing to modulate her greed by the (perceived) indignation of the responder.

Assuming that self-interest (however displayed, via the payoff structure or via an emotional state of mind) is the motivating attribute of the players, we may take *intemperance* and *indignation* as the dominant social attributes of this game (altruism is also a possible attribute, but the analysis is simpler if we assume that such tendencies are not significant). We will denote these two attributes by the *intemperance index* α and the *indignation index* β, and assume that $0 \leq \alpha \leq 1$ and $0 \leq \beta \leq 1$. The condition $\alpha \approx 1$ means that the proposer is avaricious, while $\alpha \approx 0$ means that the proposer is moderate and willing to restrain her demand for wealth. The condition $\beta \approx 1$ means that the responder is extremely prone to indignation, while $\beta \approx 0$ means that the responder is extremely tolerant.

To formulate a satisficing representation of this game, we must first establish operational definitions for the notions of selectability and rejectability. Selectability will be characterized by the fundamental goal of the game, which is to obtain a share of the fortune. Rejectability, on the other hand, will be characterized by costs or hazards which, for this game, is failure to consummate a deal. The following application of the chain rule is an appropriate way to proceed:

$$p_{S_1 S_2 R_1 R_2}(u_1, u_2; v_1, v_2) = p_{S_2|S_1 R_1 R_2}(u_2|u_1; v_1, v_2) \cdot p_{R_1|S_1 R_2}(v_1|u_1; v_2)$$
$$\cdot p_{R_2|S_1}(v_2|u_1) \cdot p_{S_1}(u_1). \tag{8.23}$$

We make the following simplifying assumptions: (a) a single player's selectability and rejectability are independent; and (b) conditioned on X_1's selectability, X_2's selectability is independent of X_1's rejectability. With these assumptions, we may simplify (8.23) to become

$$p_{S_1 S_2 R_1 R_2}(u_1, u_2; v_1, v_2) = p_{S_2|S_1}(u_2|u_1) \cdot p_{R_1|R_2}(v_1|v_2) \cdot p_{R_2|S_1}(v_2|u_1) \cdot p_{S_1}(u_1). \tag{8.24}$$

To compute $p_{S_2|S_1}$, we observe that, if X_1 were to place all of her selectability mass on h, then X_2 should place all of his selectability mass on a. But if X_1 were to place all of her selectability mass on l, then X_1 would be indignant and would select d with weight β. The resulting conditional mass functions are

$$p_{S_2|S_1}(a|h) = 1, \qquad p_{S_2|S_1}(d|h) = 0,$$
$$p_{S_2|S_1}(a|\ell) = 1 - \beta, \qquad p_{S_2|S_1}(d|\ell) = \beta.$$

To compute $p_{R_1|R_2}$, suppose that X_2 were to place all of his rejectability mass on d. If X_1 were to know that X_1 would reject declining and hence would accept the offer, the amount she would offer would depend upon her intemperance. If her intemperance index were high, most of the mass would go to rejecting h. However, in the interest of avoiding failure, X_1 would be wise to modulate her temerity with the indignation she expects X_2 to feel if a low offer were tendered. On the other hand, if X_2 were to place all of his rejectability mass on a, then he would determine to decline the offer. However, even in that situation, there is no reason for X_1 not to accommodate X_2's indignation, so she might as well continue to offer the same deference to X_2. The corresponding conditional rejectability functions for X_1 are

$$p_{R_1|R_2}(h|a) = \alpha(1-\beta), \qquad p_{R_1|R_2}(\ell|a) = 1 - \alpha(1-\beta),$$
$$p_{R_1|R_2}(h|d) = \alpha(1-\beta), \qquad p_{R_1|R_2}(\ell|d) = 1 - \alpha(1-\beta).$$

To compute $p_{R_2|S_1}$, we observe that, if X_1 were to place all of her selectability mass on h, then X_2 would surely accept the offer, hence would ascribe his entire unit of conditional rejectability mass to d. If, however, X_1 were to ascribe her entire unit of conditional rejectability mass to ℓ, then X_2 would reject a according to his indignation factor. The resulting conditional rejectability mass functions for X_2 are

$$p_{R_2|S_1}(a|h) = 0, \qquad p_{R_2|S_1}(d|h) = 1,$$
$$p_{R_2|S_1}(a|\ell) = \beta, \qquad p_{R_2|S_1}(d|\ell) = 1 - \beta.$$

Finally, to compute p_{S_1} we note that, without taking into account any other factors, X_1 would offer ℓ in proportion to her intemperance, thus,

$$p_{S_1}(h) = 1 - \alpha, \qquad p_{S_1}(\ell) = \alpha.$$

The product of these conditional probabilities composes the interdependence function which can then be used to generate the selectability and rejectability marginals that define group and individual satisficing behavior. For this problem, the product generates rather complicated expressions which are best evaluated with the aid of a computer as functions of the intemperance and indignation parameters.

Figure 8.3 illustrates plots of $p_{S_1}(h)$ and $p_{R_1}(h)$ as functions of α and β as they each vary over the interval $[0, 1]$. These two functions appear as sheets in the figure. With $q = 1$, the intersection of these sheets is the line of demarcation between X_1 choosing h (i.e., when $p_{S_1}(h) \geq p_{R_1}(h)$). Notice that when intemperance (α) is low, X_1's satisficing decision is to offer h for all values of perceived indignation (β). But, as α increases toward unity, X_1's decision to choose ℓ becomes sensitive to X_2's perceived indignation. This situation illustrates X_1's consideration of X_2's perceived attitude: the higher X_1's indignation, the higher must be X_1's intemperance in order for her to try to take advantage of X_2.

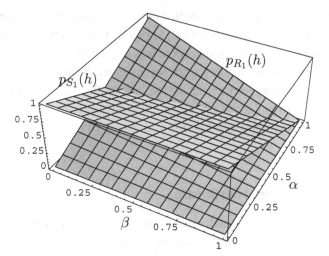

Figure 8.3: The proposer's decision rule for the satisficing Ultimatum minigame.

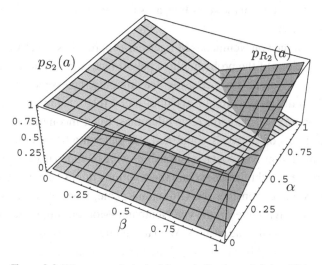

Figure 8.4: The responder's decision rule for the satisficing Ultimatum minigame.

Figure 8.4 illustrates plots of $p_{S_2}(a)$ and $p_{R_2}(a)$ as functions of α and β, as they each vary over the interval $[0, 1]$. These two functions appear as sheets in the figure. With $q = 1$, the intersection of these sheets is the line of demarcation between X_2 choosing a (i.e., when $p_{S_2}(a) \geq p_{R_2}(a)$). For low to moderately high values of either X_1's perceived intemperance index or X_2's indignation index, X_2's satisficing decision is to accept the offer. But, if X_1 is perceived to be extremely intemperate and X_2 is extremely indignant, then X_2's satisficing decision is to decline the offer.

The satisficing formulation of the Ultimatum minigame requires each player to determine the other player's social index. This may be difficult, especially for the proposer.

Although she has the advantage of making the first move, she must rely upon an *a priori* estimate of X_2's indignation. Fortunately for her, as Figure 8.4 illustrates, her choice is not sensitive to the indignation of the responder unless she is extremely intemperate.

On the other hand, since the responder has the second move, he has the advantage of estimating the proposer's intemperance from the size of her offer. He may set his value of α to be inversely proportional to the size of the offer, using that estimate along with his indignation index to position himself on the grid displayed in Figure 8.4 to make his decision.

8.1.5   The game-theoretic role of social relationships

The satisficing solutions to the four games discussed above have a common property, namely, the solutions are parameterized by indices that characterize the social dispositions of the players, and the decisions change as the parameters are varied according to these dispositions. Even though the games are expressed in mathematical form, they are not completely abstracted from the story line, as is the case with the von Neumann–Morgenstern representation.

One of the strengths of the von Neumann–Morgenstern approach is that, by encoding the important considerations of the problem into a payoff matrix, all irrelevant aspects of the problem (the "fat") can be stripped away, leaving only the essence. But, if the simplifications result in the elimination of essential social relationships, the resulting game may generate anomalies, paradoxes, or dilemmas. Such incongruities may be nothing more than artifacts of an overly simplified and under-modeled representation of the original problem. If, however, social relationships were included in the game, then the apparant incongruities might disappear and the solutions might be consistent with common sense. For example, consider the Ultimatum game, where (as the evidence with respect to human behavior attests) the von Neumann–Morgenstern representation fails to consider essential social relationships. With this game, a good case can be made that the "anomaly" of irrational behavior is nothing more than an artifact.

Since satisficing game theory is not based on individual rationality, social relationships can be relevant and there is no need to strip them away. Satisficing game theory does not *require* the introduction of social relationships, but it does accommodate them. If no such relationships exit, then the interdependence function can encode exactly the same information as is encoded in a payoff matrix, and the satisficing game will give the von Neumann–Morgenstern solution. For example, with the Prisoner's Dilemma, setting the vulnerability index to zero and the dissociation index to unity (and thereby eliminating any social relationships from the problem) results in the Nash equilibrium of mutual defection.

As Rasmusen correctly observes, expanding from a simplified problem to the original problem is perilous. It is especially perilous for artificial decision-making entities.

Humans may be able to compensate for anomalies and paradoxes by examining them in the context of the original problem, but it is doubtful that machines will be able to do so successfully. Machines are incapable of either rational or irrational behavior. They are also incapable of recognizing anomalies, paradoxes, and dilemmas. They do exactly what they are programmed to do, and oversimplified machine logic will lead to dysfunctional performance.

Since individual rationality obviates social relationships and therefore can result in oversimplified models of social behavior, it is not an adequate paradigm for developing models that possess the social sophistication needed for functional artificial societies in cooperative environments. Machines that are expected to perform in complex social environments must possess commensurately complex models of the society. Thus, rather on concentrating on the removal of "fat" in the interest of simplification, the more important issue is how to introduce, coherently, unambiguously, and efficiently, the complexity that will be necessarily present in any functional society of artificial decision makers. Satisficing game theory is one way to address this issue.

8.2   Mitigating complexity

The above examples illustrate that, if we allow the preferences of a decision maker to be influenced by the preferences of other decision makers, the complexity of the system increases at an exponential rate with the number of decision makers. It is vital that this complexity be managed properly and that system models do not degenerate into a messy tangle of arbitrary linkages. We must insist upon three properties of such a model. It must be self-consistent, in that none of the preference linkages are contradictory. It must be complete, in that all relevant linkages are accommodated. Finally, it must be non-redundant. Fortunately, the representation of these linkages via the structure of probability theory accommodates these three properties in a parsimonious and fairly facile way.

A contradiction arises if a given premise leads to a conclusion and its negation. If the sky is cloudy, we may conclude that the probability of rain is increased. It would be a contradiction to assert that the probability that it will not rain is also increased. Fortunately, the mathematical structure of probability does not permit such contradictory assertions. Probabilistic conclusions are matters of degree, and it is the degree of the assertion that is guaranteed to be non-contradictory. Since interdependency functions have the mathematical structure of probability, they inherit the property of non-contradictoriness. With the Battle of the Sexes game, for example, conditional selectabilities such as those given by (8.11) ensure that the relationships between H and S cannot be contradictory. Given that S rejects D, H commits to select B, and not to select D. Thus, H cannot have contradictory preferences.

The interdependence function provides a complete description of all preference relationships. There is no preference linkage between decision makers that cannot be
captured by the appropriate choice of conditional interdependence. Although in practice many of the links between participants may be null, the links nevertheless exist
and may be activated if desired. The interdependence function possesses no redun-
dancy. The removal of any link joining any two participants destroys the direct linkage,
and any path through other participants cannot capture the isolated behavior that was
eliminated.

The interdependence function permits full selectability and rejectability connectivity
between all participants, where each participant influences, and is influenced by, every
other member of the system. Such a fully-linked system is maximally complex. At
the other extreme, each participant may be designed to function myopically, with total
disregard for all other participants. Between these completely coupled and completely
decoupled extremes, however, are many useful structures which, if properly exploited,
will permit the functional design of multi-agent systems that are capable of coordinated
behavior. Many systems of interest are only sparsely connected, such that any given
agent influences or is influenced by only a small subset of the entire population. Such
systems exhibit a form of spatial localization. To illustrate, consider two models of
intermediate complexity.

The first is a simple hierarchical model that would be appropriate for decision makers
who reside in a military-like society or in any society where decision makers and
resources have priorities but rank takes precedence. The second model involves decision
makers whose actions must coordinate with their nearest neighbors, but, conditioned
on what their immediate neighbors do, are independent of more distant participants. Let
us refer to this as a Markovian model, since the interdependence function is assumed
to possess the Markov property of conditional independence.

Hierarchical

A system $\mathbf{X} = \{X_1, \ldots, X_N\}$ of decision makers is a **linear hierarchy** if there exists a
rank-ordering function, r, such that $r(X_i) > r(X_j)$ means that X_j must wait until X_i has
made its choice before X_j can choose. Without loss of generality, we assume that the decision makers are indexed in rank-decreasing order, that is, $r(X_1) > \cdots > r(X_N)$. The
decision-making process is initialized by X_1 computing a satisficing decision and com-
municating this information to X_2, who computes its satisficing decision conditioned
on X_1's choice and communicates both choices to X_3. X_3 performs a similar operation,
and the process continues until X_N makes its satisficing decision conditioned on the
choices of all participants higher up in the hierarchy. This approach takes full advantage
of the hierarchy and ensures that all participants will make conditionally satisficing decisions. The decision maker at the top of the hierarchy may restrict its attention to its
own self-interest, thereby simplifying the linkages between itself and lower-ranking decision makers. Similarly, as we move down the hierarchical chain, subsequent decision

makers may do likewise with respect to the agents below them in the hierarchy. This structure is applicable to a number of practical applications, such as leader–follower architectures for robotic systems, satellite formations, and military-like operations.

Markovian

Suppose that the members of a decision-making system are associated through some type of functional or proximal relationship, such as being neighbors, and that each decision maker has direct communication with and awareness of only these neighbors. For example, consider a array of mobile robots such that each robot is aware only of the other robots within its limited field of view. Let $\mathbf{X} = \{X_1, \ldots, X_N\}$ be such a system and let $\mathcal{F}_i(t)$ be the subset of $\mathbf{X}$ that lie in X_i's field of view at time t. The interdependence function then factors into the form (ignoring arguments):

$$p_{S_1 \ldots S_N R_1 \ldots R_N} = \prod_{i=1}^{N} p_{S_i R_i | \mathcal{F}_i},$$

where $p_{S_i R_i | \mathcal{F}_i}$ is the conditional interdependence of X_i with the other decision makers in its field of view. This local interdependence represents all of the information that is in the possession of X_i at time t, and thus is the only information that can be used by X_i for negotiated decision making. As the scenario evolves, however, participants may enter or leave a given field of view. This type of non-stationary, open environment is accommodated by this Markovian structure.

A multi-agent system whose interdependence function can be specified in terms of a small number of parameters is said to be *parsimonious*. Note that, even though the four games described above are quite densely interconnected, the 16 values of each of the interdependence functions are completely specified by a small set of parameters. In general, even though the linkages between decision makers may be dense, it may be possible to express these linkages in terms of only a few fundamental parameters.

Furthermore, as these examples indicate, the conditional linkages between decision makers are often expressed in terms of zero or 1. This structure indicates a very decisive relationship between participants and greatly simplifies the specification. Such relationships are very much like behavioral production rules. For example, the conditional selectability function given by (8.11) is an exact mathematization of the natural language rule

If S rejects D, then H must select B.

If S rejects B, then H must select D.

The correspondence between rules in a knowledge base and the interdependency function is a potentially very useful one. For many applications, the strength of these preference linkages must be determined by human judgment. Such judgment often takes the form of parametric representations such as those illustrated by the above examples. When such judgments are possible, the structure of the factors that compose the interdependence function may be greatly simplified.

8.3   An *N*-player example

Consider a system of decision makers $\mathbf{X} = \{X_1, \ldots, X_N\}$ whose members are collectively tasked to form themselves into an equally spaced column and proceed at a known collective velocity in this formation. To make the problem a little more interesting, let us suppose that the spacing between participants is dynamic; that is, it changes over time according to some function that is provided to all participants. We will assume, however, that each participant has a limited field of view and can sense only the participants immediately in front of and to the rear of itself. It has no knowledge of any other participants (for example, the lead participant senses only the participant to its rear, and the rearmost participant senses only the participant in front of it). This is the Markovian assumption. Suppose time flows in discrete increments, that is, $t \in \{0, 1, 2, \ldots\}$, and that at each t each participant is able to determine the distances between itself and and its immediate neighbors. Let $x_i(t)$ denote the position of X_i at time t; then each interior participant X_i, $i = 2, 3, \ldots, N - 1$, knows $y_{i,i+1}(t) = x_{i+1}(t) - x_i(t)$ and $y_{i-1,i}(t) = x_i(t) - x_{i-1}(t)$ for each t, with the front and rear participants knowing $y_{1,2}(t) = x_2(t) - x_1(t)$ and $y_{N-1,N}(t) = x_N(t) - x_{N-1}(t)$, respectively. Let $v(t)$ denote the collective velocity and let $c(t)$ denote the desired spacing between participants. Each participant must determine how to change its velocity to conform to the equal-spacing requirement. Let $u_i(t)$ denote the discrete position change command that X_i makes in response to its environment, and assume that the set of possible commands is the set $\mathbf{U} = \{\mu_1, \ldots \mu_n\}$. For this problem we shall take $\mu_k = d(k - 2)$ and $n = 3$, yielding the action space, $\mathbf{U} = \{-d, 0, d\}$.

Since each interior participant is aware of only itself and its immediate neighbors, we may form a family of $N - 2$ linear systems of the form

$$x_{i-1}(t + 1) = x_{i-1}(t) + u_{i-1}(t),$$
$$x_i(t + 1) = x_i(t) + u_i(t),$$
$$x_{i+1}(t + 1) = x_{i+1}(t) + u_{i+1}(t),$$
$$y_{i-1,i}(t) = x_i(t) - x_{i-1}(t),$$
$$y_{i,i+1}(t) = x_{i+1}(t) - x_i(t),$$

for $i = 2, \ldots, N - 1$, with

$$x_1(t + 1) = x_1(t) + u_1(t),$$
$$x_2(t + 1) = x_2(t) + u_2(t),$$
$$y_{1,2}(t) = x_2(t) - x_1(t),$$

and

$$x_{N-1}(t+1) = x_{N-1}(t) + u_{N-1}(t),$$
$$x_N(t+1) = x_N(t) + u_N(t),$$
$$y_{N-1,N}(t) = x_N(t) - x_{N-1}(t),$$

for the first and last members of the platoon. Under the Markovian assumption, these linear subsystems must be controlled independently of each other. Unfortunately, none of these subsystems is observable, that is, the state cannot be uniquely reconstructed from the observations. Consequently, standard feedback control techniques cannot be applied. This is a coordination game of asymmetric information, which we dub the *Markovian Platoon* game. A distinguishing characteristic of this game is that no proper subset of the set of participants possesses all of the information necessary to make a decision for the entire set. We approach this problem by first obtaining an optimal solution. We then apply the satisficing theory by computing a Markovian interdependence function for each participant.

8.3.1   The optimal solution

We may formulate this as a optimal control problem under conditions of uncertainty; that is, each participant attempts to optimize its performance given whatever knowledge it possesses while accounting as best it can for knowledge limitations. Let us consider an interior participant, say X_i, $i \in \{3, \ldots, N-2\}$. X_i knows the distances between X_i and X_{i+1} and between X_i and X_{i-1}, but does not know the distances between either X_{i+1} and X_{i+2}, or X_{i-2} and X_{i-1}. Consequently, X_i cannot know with certainty how either X_{i+1} or X_{i-1} will respond. We adopt the following performance index for X_i:

$$J_i(u_i, u_{i-1}, u_{i+1}) = ||y_{i-1,i} + u_i - u_{i-1}| - c| + ||y_{i,i+1} + u_{i+1} - u_i| - c|.$$

Unfortunately, X_i cannot simply minimize this function with respect to its command, u_i, since it does not know what commands X_{i-1} and X_{i+1} will make. Under the optimal approach, X_i should choose the value for u_i that minimizes the expected value of J_i, where the expectation is taken with respect to a joint probability characterizing the behavior, from the point of view of X_i of the joint decision (u_{i-1}, u_{i+1}). The guiding principle for the formulation of such a joint probability is that, if the distance between X_i and X_{i-1} is less than c, then, from the point of view of X_i (i.e., without any knowledge of X_{i-2}), X_{i-1} would be more likely to move away from X_i than toward X_i, with a similar relationship holding between X_i and X_{i+1}. This principle may be implemented with the following mass function:

$$q_{i-1,i+1}(u_{i-1}, u_{i+1}|y_{i-1,i}, y_{i,i+1}) \propto \frac{1}{||y_{i-1,i} - u_{i-1}| + |y_{i,i+1} + u_{i+1}| - c| + \epsilon}$$

for $i = 2, \ldots, N-1$, where ϵ is a small quantity added to ensure that the ratio is bounded. To form its decision, X_i then computes

$$u_i^* = \arg\min_{u \in U} \sum_{j=1}^n \sum_{k=1}^n J_i(u, \mu_j, \mu_k) q_{i-1,i+1}(\mu_j, \mu_k | y_{i-1,i}, y_{i,i+1})$$

for $i = 2, \ldots, N - 1$. The boundary participants, X_1 and X_N, implement

$$u_1^* = \arg\min_{u \in U} \sum_{j=1}^n J_1(u, \mu_j) q_2(\mu_j | y_{1,2}),$$

$$u_N^* = \arg\min_{u \in U} \sum_{j=1}^n J_N(u, \mu_j) q_{N-1}(\mu_j | y_{N-1,N}),$$

where

$$J_1(u_1, u_2) = ||y_{1,2} + u_2| - c|,$$
$$J_N(u_{N-1}, u_N) = ||y_{N-1,N} + u_{N-1}| - c|,$$

and

$$q_2(u | y_{1,2}) \propto \frac{1}{|y_{1,2} + u_2| + \epsilon},$$

$$q_{N-1}(u | y_{N-1,N}) \propto \frac{1}{|y_{N-1,N} - u_{N-1}| + \epsilon}.$$

Figure 8.5 illustrates the results of a simulation involving $N = 6$ participants with time along the horizontal axis and the relative position of the participants plotted along the vertical axis. Note that the group velocity as been subtracted from each individual velocity, resulting in "in place" marching. Each column of symbols represents a snapshot of the participants at the given time. For this simulation,

$$c(t) = 1 + 0.05 \sin(2\pi t / 70),$$
$$\delta = 0.05,$$

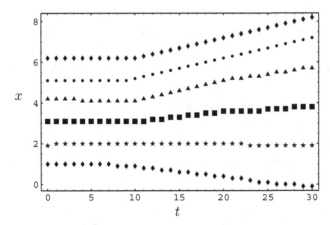

Figure 8.5: The optimal solution to the Markovian Platoon.

and the initial spacings between participants was chosen pseudo-randomly according to a uniform distribution over a range of ± 0.15 centered at the desired spacing. As time evolves, the desired spacing between the participants is modulated at a low frequency – this time-varying spacing is assumed as common knowledge for all participants. This simulation indicates that the optimal solution is basically stable, yielding an RMS (root mean square) error of 0.621 averaged over all participants over 30 time increments over 50 independent trials.

8.3.2   The satisficing solution

To formulate a satisficing solution, we must first establish operational definitions for selectability and rejectability. The selectability of an action will be determined by the degree to which it achieves the desired spacing, and the rejectability of an action will be taken as the degree to which the participants are apt to collide with each other. We will assume a Markovian structure, yielding a local interdependence function for each interior participant of the form

$$p_{S_{i-1} S_i S_{i+1} R_{i-1} R_i R_{i+1}} = p_{S_i | S_{i-1} S_{i+1} R_{i-1} R_i R_{i+1}}$$
$$\cdot\, p_{S_{i-1} S_{i+1} | R_{i-1} R_i R_{i+1}} \cdot p_{R_i | R_{i-1} R_{i+1}} \cdot p_{R_{i-1} R_{i+1}}, \qquad (8.25)$$

for $i = 2, \ldots, N - 1$. For X_1 and X_N, the local interdependence functions are

$$p_{S_1 S_2 R_1 R_2} = p_{S_1 S_2 | R_1 R_2} \cdot p_{R_1 R_2}, \qquad (8.26)$$
$$p_{S_{N-1} S_N R_{N-1} R_N} = p_{S_{N-1} S_N | R_{N-1} R_N} \cdot p_{R_{N-1} R_N}. \qquad (8.27)$$

Using the above-defined working definitions for selectability and rejectability, we may form the factors of (8.25). Let us consider the first factor, $p_{S_i | S_{i-1} S_{i+1} R_{i-1} R_i R_{i+1}}$. Employing a functional form similar to what was used for the optimal solution, we express this conditional selectability mass function as

$$p_{S_i | S_{i-1} S_{i+1} R_{i-1} R_i R_{i+1}}(u_i | u_{i-1}, u_{i+1}; v_{i-1}, v_i, v_{i+1}) \propto \frac{1}{A + B + \epsilon},$$

where

$$A = (|y_{i-1,i} + v_i - v_{i-1}| + |y_{i,i+1} + v_{i+1} - v_i|) \cdot (||y_{i-1,i} + u_i - u_{i-1}| - c|)$$

and

$$B = ||y_{i,i+1} + u_{i+1} - u_i| - c|.$$

By a similar argument, we may express the second factor of (8.25) as

$$p_{S_{i-1} S_{i+1} | R_{i-1} R_i R_{i+1}}(u_{i-1}, u_{i+1} | v_{i-1}, v_i, v_{i+1}) \propto \frac{1}{C + D + \epsilon},$$

where

$$C = (|y_{i-1,i} - v_{i-1}| + |y_{i,i+1} + v_{i+1}|) \cdot (||y_{i-1,i} - u_{i-1}| - c|)$$

and

$$D = ||y_{i,i+1} + u_{i+1}| - c|.$$

We may complete the specification of (8.25) by computing the third and fourth factors as

$$p_{R_i|R_{i-1}R_{i+1}}(v_i|v_{i-1}, v_{i+1}) \propto \frac{1}{|y_{i-1,i} + v_i - v_{i-1}| + |y_{i,i+1} + v_{i+1} - v_i| + \epsilon},$$

$$p_{R_{i-1}R_iR_{i+1}}(v_{i-1}, v_i, v_{i+1}) \propto \frac{1}{|y_{i-1,i} - v_{i-1}| + ||y_{i,i+1} + v_{i+1}| + \epsilon}.$$

For X_1 and X_N, the factors of (8.26) become

$$p_{S_1S_2|R_1R_2}(u_1|u_2, v_1, v_2) \propto \frac{1}{|y_{1,2} + v_2 - v_1| \cdot ||y_{1,2} + u_2 - u_1| - c| + \epsilon},$$

$$p_{R_1R_2}(v_1, v_2) \propto \frac{1}{|y_{1,2} + u_2 - u_1| + \epsilon},$$

$$p_{S_{N-1}S_N|R_{N-1}R_N}(u_{N-1}, u_N|v_{N-1}, v_N)$$
$$\propto \frac{1}{|y_{N-1,N} + v_N - v_{N-1}| \cdot ||y_{N-1,N} + u_N - u_{N-1}| - c| + \epsilon},$$

$$p_{R_{N-1}R_N}(v_{N-1}, v_N) \propto \frac{1}{|y_{N-1,N} + u_N - u_{N-1}| + \epsilon}.$$

Figure 8.6 illustrates the satisficing solution to the Markovian Platoon problem obtained by adopting the maximally discriminating satisficing decision. These results also

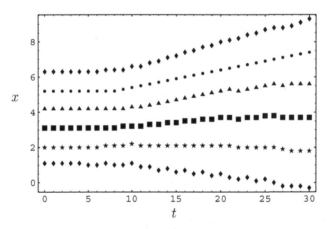

Figure 8.6: The satisficing solution to the Markovian Platoon.

demonstrate stability. The satisficing solution tracks essentially the same as does the optimal solution, with an RMS error of 0.615 averaged over all participants over 30 time increments over 50 independent trials.

Both optimization theory and satisficing theory provide systematic synthesis procedures, and they both have claims to being rational. They differ in the way goals and costs are expressed and in the decision mechanism used. With the optimal approach, knowledge is expressed asymmetrically via a performance index and a probabilistic description of uncertainty, and decisions are made by making intra-utility comparisons. With the satisficing approach, knowledge is expressed symmetrically via the interdependence function, and decisions are made by making inter-utility comparisons. These representations, though different in structure, may be made to be essentially equivalent (as evidenced in the Markovian Platoon problem) by considering the same attributes, employing compatible assumptions, and by quantifying the significance of attributes in similar units.

9 Meliority

The protest of common sense [is] the complaint of mere habit.

Michael Polanyi
Personal Knowledge (University of Chicago Press, 1962)

Principles of rationality serve as normative models of behavior, but rationality can never be anything more than an ideal. Even the most determined adherent to a particular code of rationality must recognize that circumstances often prohibit the attainment of the ideal in practice. This realization, however, should not change the standard that is sought. Indeed, failure to achieve the standard should serve as motivation to improve the ability to perform, rather than a rationale for lowering the standards.

Perhaps the most salient criticism of satisficing *à la* Simon's aspiration levels (see, e.g., Levi (1997) and Zeleny (1982)) and other forms of 'bounded rationality' is that they condone the deliberate acceptance of a solution that is less than the ideal. In terms of human behavior, deliberately falling short of achieving the goal to which one has made a commitment may be disingenuous,[1] but at least the decision maker can mount an explanation for his or her failure. One possible explanation for a decision maker who has apparently committed to optimality only to equivocate is that, after all, he or she is not really an optimizer, but is merely using the tools of optimization to find a solution that is good enough. If it happens to be optimal, so much the better, but that would be a bonus, not a necessity.

The trouble is, that while humans are quite capable of operating with a double standard, it is not clear how to endow an artificial decision-making entity with such an ability. Equivocation may be as difficult a concept to capture in a machine as is any other human attribute. In those circumstances where being good enough, and not optimality, is the real goal, then it is imperative that a clear definition of what it means to be good enough be made explicit, and that decision making under this definition be removed from the realm of *ad hoc* activity.

---

[1] Levi condemns as fraudulent the making of a promise that one is certain will not be kept, and argues that one is obligated to seek training, education, and the use of appropriate therapies and prosthetic devices to enhance capabilities in a sincere attempt to achieve the ideal.

9.1   Amelioration versus optimization

For many decades, substantive rationality has been the paradigm of choice with which to address decision problems. Much effort has been devoted to the art of making rational choices according to various notions of optimization. Substantive rationality has more than proven its worth in countless applications – success has been virtually unqualified. In the engineering domain, control theory, estimation and detection theory, communication theory, and various signal processing techniques are all instantiations of automated decision making, with similar success stories in computer science, statistics, and economics. The foundational principles upon which these applications rest is almost always some manifestation of optimization.

This book is an invitation to consider amelioration, as well as optimization, as a species of rational decision making. Decision makers will need good reasons to consider a change from a convention that is as compelling as optimization. The old adage, "If it's not broken, don't fix it," holds considerable power. To make any claims that the method presented here should supplant existing approaches that have enjoyed demonstrated success would be inappropriate without compelling evidence. I do not recommend the general redoing of more than a half-century of formalized decision-making theory.

At the same time, I point out that amelioration does not contradict optimization. Rather, it softens it. I do not abandon optimization, but I do consider amelioration as an alternative. It should be considered on its own merits as appropriate for a given application and may even be more appropriate than optimization.

The art of decision making is in a state of continual change as new problems are considered. Dramatic increases in computational power make it possible to address problems of ever increasing complexity. One of the great success stories of computing is the automation of decision making. Computers can be much more than prostheses to facilitate computation. They can also be prostheses of the mind and serve as instruments with which to explore expanded domains. One of these domains is the progression from merely automated (self-acting) decision making to truly autonomous (self-governing) decision making. To achieve this capability, however, it is likely that many refinements and alternatives to conventional approaches must be explored. Some of these alternatives may not have the apparent power, efficiency, and security that conventional approaches afford, but they may have compensating attributes that make their consideration worthwhile.

There are a number of properties of satisficing decision theory that can be exploited to advantage by both single and multiple decision makers. These properties are:

(a) binary choices based on local criteria,

(b) multiple solutions, and

(c) guiding versus dictating solutions.

In multiple-agent decision-making scenarios, some additional properties may be exploited, including

(d) accommodation of group interest,

(e) mitigation of competition when it is not natural, and

(f) flexible negotiation.

Satisficing requires making binary decisions, where the decision is either to retain consideration of an option as good enough for adoption or to reject it as not good enough. This is perhaps the most primitive form of decision making and does not involve comparisons with other options. Instead, it is a local decision, based only on the intrinsic merits of the option. Options that survive this binary comparison may be viewed as being "better" options, in the sense that the benefits are "better" than the costs, but they may not be viewed as being best in any reasonable sense.

In general, satisficing yields a set of solutions rather than just one solution. Since every member of this set has the property that the gains at least compensate for the losses, implementing any of them is ameliorative with respect to choosing an option not in the satisficing set. The set can be further refined in a number of ways. One way is to impose the additional requirement of equilibrium by eliminating those satisficing options that cost more but do not provide more benefit or that provide less benefit but do not cost more. Another way to refine this set is to eliminate those options that are not securely satisficing, that is, that do not meet or exceed an externally applied threshold. A third way to prune the satisficing set is to eliminate satisficing options whose benefits do not exceed a fixed benefit threshold or whose costs do exceed a fixed cost threshold. These systematic ways of refining the satisficing set, however, may still leave the decision maker with multiple options. Simply choosing one at random would perhaps be acceptable, but, at this point, it might also be reasonable to resort to substantive rationality and pick a single option in the set according to an ancillary criterion.

The selectability and rejectability criteria are not designed to identify a single option but only to identify options that are defensibly good enough. *Since the decision maker has some ultimate flexibility in making the final choice, the criteria do not absolutely dictate the solution, as would be the case with optimality. By mitigating the demand that the best and nothing but the best be selected, the decision maker is provided with a set of selections, each of which enjoys a claim to being adequate. This flexibility permits the decision maker to accommodate its own non-quantifiable biases, should it desire, in making the ultimate decision.*

Perhaps the most important distinction between satisficing and approaches based ultimately on a notion of optimality occurs in multiple decision-maker environments. In multi-agent settings, optimality is usually instantiated via von Neumann–Morgenstern game theory, but that theory is based on the notion of exclusive self-interest. Each decision maker has its own utility function which, although of necessity is a function of the options available to all players of the game, is not a function of the preferences of the other players. It is not possible to create a notion of group interest that is based on exclusive self-interest. Satisficing, however, does provide an avenue to account for group interest, since the interdependence function accounts for couplings between decision-maker preferences.

The ultimate consequence of group interaction where each decision maker is focused primarily on its own self-interest is the engendering of competition. Even if the decision makers are not naturally disposed to compete, competition can arise as an artifact of each decision maker attempting to maximize its own benefits. Sharing cannot be accommodated if doing so would put any of the decision makers at a voluntary disadvantage, no matter how slight the disadvantage may be or how beneficial sharing would be to others. A related consequence of competition is the creation of artificial scarcity even if there is seemingly an abundance of resource, since it may be possible (and therefore mandatory under maximization) for some decision makers to accumulate more resources than they need. Satisficing, on the other hand, does not engender competition unless it naturally arises due to conflicting preferences of the decision makers. Furthermore, since the decision makers are not mandated to maximize, they are able to participate in sharing strategies that can accommodate the interests of others, even if it is at their own expense.

Negotiation is especially problematic under substantive rationality. Negotiation usually requires compromise, but substantive rationality brooks no compromises that do not guarantee the decision maker at least as much as it could obtain if it did not participate in negotiations. When negotiation is constrained by optimization, there may be no joint strategy that is acceptable for all decision makers. Under satisficing, however, it is possible to devise negotiation strategies that provide all decision makers with options that are good enough for themselves and with the joint options that are good enough for the group as a whole.

With all of these potential advantages of satisficing, however, there are some disadvantages which, if not dealt with properly, could undermine the satisficing approach. First, the fact that decisions are made according to local binary criteria opens the door to accepting decisions whose primary virtue is that that they are cheap. Both an inexpensive economy car or a moderate expensive luxury sedan may satisfy a buyer's satisficing equation, but the buyer may not be neutral between the choices. The notions of dubiety, gratification, and ambivalence, however, serve to further classify such decisions, and provide the decision maker with additional criteria by which to make a final choice.

Perhaps the most challenging concern with satisficing, especially in a multi-agent environment, is how to deal with the greater complexity compared with conventional game theory in accounting for other decision makers' preferences. The growth of this kind of complexity can sometimes be controlled by the use of various kinds of localization assumptions, but, ultimately, I have no sure-fire way to keep the complexity at a manageable level. My only defense of this situation is that, complex though the models may be, they are not more complex than they need to be to characterize the essential inter-relationships that exist between decision makers. Complexity is undoubtedly the biggest problem that automated multi-agent decision making will face in the foreseeable future. As far as constructing realistic models to account for the possible interactions between decision makers is concerned, satisficing theory seems no worse off and no better off than any other methodology.

9.2   Meta-decisions

One of the most important issues that a decision maker or the designer of an artificial decision-making system must face is the meta-decision of choosing the decision-making paradigm. Most decision-making approaches appearing in the literature are based on two major paradigms: substantive rationality and procedural rationality. If these are the only paradigms available, then the meta-decision itself must be restricted to these two paradigms. Suppose one were to approach the meta-decision from the perspective of substantive rationality. One would then choose the paradigm with the larger expected utility. On the other hand, if one were to approach the meta-decision problem from the perspective of procedural rationality, one would then apply a heuristic that would dictate the choice. But how does one define either a utility or a heuristic by which to compare paradigms? How does one avoid the infinite regress problem in either case? Is it even possible, when forming such a decision problem, to distinguish between a heuristic and a utility function? Even more fundamentally, is it even possible to make a meaningful comparison between the two paradigms? These questions seem to be undecidable.

One way to address these questions is to assert that it is possible to proceed in a rational way without even asking them, let alone answering them. But doing so requires that a third choice mechanism be invoked, such as intrinsic rationality. *Intrinsic rationality permits the decision maker to consider each of the paradigms on its own merits without the need to compare them.* If, in the deep-seated, almost instinctive, judgment of the paradigm chooser, the benefits that would be derived by adopting substantive rationality outweigh the costs, then substantive rationality is to be seriously considered – it is good enough, or satisficing. Also, procedural rationality can be evaluated as either satisficing or not. But, since these two paradigms are no longer the only possible mechanism for making choices, one may also consider the satisficing status of intrinsic rationality itself as the paradigm of choice.

The choice of which notion of rationality is to be observed when designing artificial decision-making systems cannot be avoided. The choice will in large part determine the operational characteristics, or "personality," of the system. A goal of this book is to place intrinsic rationality on the table as being worthy of serious consideration, both for analyzing existing decision-making activities and as a means for designing artificial decision-making systems.

To illustrate, consider the application domain of integrating technology and human performance in the design of automated systems to enhance safety and performance in automobile driving. The study of safety augmentation systems for collision and accident avoidance and performance augmentation systems for adaptive speed control represent important, active research areas that involve decision making in critical situations. Such systems must be designed in a way that is compatible with human performance. In fact, if a system does not perform in a way that is similar to the way the human operator performs, the operator will fail to have confidence in it, no matter how state-of-the-art and reliable it may be, and will not use it.

Human automobile drivers are seldom optimizers. A driver on a winding road will navigate curves no faster than he or she deems safe. Whereas an experienced race-car driver may be able to define the fastest safe speed and successfully navigate a turn at that speed, such performance would never be attempted by an average driver. Even average drivers will differ significantly in what they regard as the safe speed to navigate any given curve. When traveling on the open road, an optimizer would insist on traveling at exactly the speed limit, but the evidence clearly indicates a great variety in the speeds of various drivers. These, and many other examples that can easily be cited, are strong indications that optimality, as a principle, is simply not appropriate for most operators of automobiles. An automobile driving augmentation system that is strongly based on some notion of optimality, therefore, is likely to be received with skepticism by many, if not most, drivers.

Driving involves making tradeoffs. It involves making comparisons between the benefits, such as speed and enjoyment, versus the costs of danger and fuel consumption, to name but a few of the considerations. A good case can be made that satisficing is a more natural paradigm for the design of, say, an adaptive cruise control device, than optimization would be. The ultimate test of the success of any such automated device is, first, if users are willing to trust their safety to it, and, second, if its operational characteristics are consistent with the operator's driving skills and other driving-related behavior.

This example illustrates that the choice of decision-making mechanism is far from obvious. Such meta-decisions will become increasingly important as technological advances provide increasing opportunities for the design of such systems. In just a few short years, computers have advanced from the laboratory to the desk-top, from the desk-top to the lap-top, and from the lap-top to the hand-held, with no end in sight as to their ultimate portability and capability. Along with this development comes the obligation to advance the ways in which they may be used and trusted as their influence permeates deeper and deeper into society.

9.3   Some open questions

Satisficing game and decision theory, as developed in this book, is a new concept and has not yet been established in operational settings as a viable decision-making paradigm. However, work to that end is progressing on several fronts. It is not the intent of this book, however, to demonstrate operational performance. My goal is much more modest: I seek to build the case for further exploration of the merits of this different point of view.

There are a number of unaddressed and unsolved problems pertaining to this new theory. The following list of open questions may provide points of departure for additional research.

1. The approach developed in this book provides a contrast to conventional game theory, which tells us about outcomes we can expect from substantively rational decision makers. Game theory has been used extensively as a means of modeling human behavior, but there is considerable evidence that people often do not behave in ways consistent with substantive rationality; that is, they are not optimizers, or even constrained optimizers (Mansbridge, 1990a; Sober and Wilson, 1998; Bazerman, 1983; Bazerman and Neale, 1992; Rapoport and Orwant, 1962; Slote, 1989). An important, open question is whether or not, and under what conditions, the notion of satisficing based on intrinsic rationality provides a valid model for human behavior; that is, are people satisficers as we have defined the term? An answer to this question may be provided by appropriately designed psychological testing and evaluation.

2. A characteristic of conventional game theory is that it employs rationality postulates that are imposed at the individual level rather than at that of the group. Game theory is designed to ensure that each participant achieves the best result possible for itself, even if it is only in the maximin sense, regardless of the effect on others. Game theory does not easily accommodate group interest, since the preferences of each decision maker are expressed as functions only of the choices of other participants and are not conditioned on their preferences. With conventional game theory, attitudes such as cooperation and conflict are not built into the utility functions. Instead, they become evident only when the utilities are juxtaposed – the linkages are external. A characteristic of our approach, however, is that preference relationships between participants can be expressed via the interdependence function – the linkages are internal, and group preferences can emerge from these linkages. This feature invites further investigation. Does the explicit linkage of inter-agent preferences provide a basis on which to construct an artificial society that captures important aspects of human social behavior?

3. Satisficing decision theory may provide an appropriate framework for the design of negotiatory processes. One of the problems with von Neumann–Morgenstern game

theory as a framework for negotiation is that it is not constructive – it may identify a best compromise but does not provide a procedure for reaching it. The main trouble is dealing with the dynamic nature of coalition formation. Consequently, the strategic form is used extensively, and much of classical game-theoretic attention has been focused on situations where the process of negotiation can be presumed irrelevant to actual play. Heuristic approaches to negotiation are amenable to the development of processes, but they lack the capacity for self-policing, and quality cannot be assured. The principal run-time "decision-making" activity under substantive rationality is searching for an option that possesses the externally defined attribute of "optimality" and, under procedural rationality, is rule following. But under intrinsic rationality, the principal run-time activity is evaluating dichotomies and making choices dynamically and interactively according to internal assessments of both group and individual preferences. This feature is potentially a great advantage when designing negotiatory processes. When negotiating, is seeking a good enough compromise a more robust and flexible posture than directly seeking a best compromise?

4. No realistic decision problem can account for all logically possible options. All decision problems are framed against a background of knowledge and assumptions that result in a subset of options that are deemed relevant by the decision maker. This set may or may not be adequate for the task at hand, and one of the most difficult of all decision-theoretic issues is to decide whether or not this set of options should be enlarged and, if so, how to go about expanding it. Rank-ordering-based techniques, by their very nature, provide only relative information and cannot be used to address this concern. Dichotomy-based techniques, such as praxeic utility theory, may stand a better chance of addressing this issue, since they are grounded in the fundamental properties of the options and permit self-policing.

For example, being unable to find a good enough option in a situation may lead a decision maker to reconsider what it is willing or unwilling to do. If, for another example, there are no options for which the selectability-to-rejectability ratio provides a clear choice, this is evidence that the decision problem is a tense one for the decision maker. This is not to say that the decision maker cannot make a good decision. Rather, it is merely evidence that it may not be well-suited, in an ecological sense, for the task. This realization may trigger the expansion of the set of options. In practical situations this may require activating additional sensors, applying more computational power, interrogating information sources, or initiating other means of acquiring additional information, perhaps at a cost, in an attempt to equip the decision maker to deal better with its environment.

5. Biologically and cognitively inspired metaphors, such as neural networks, evolutionary systems, fuzzy logic, and expert systems, though perhaps lending great anthropomorphic insight, deal primarily with the way knowledge is encoded and represented. They do not offer new logics for decision-making processes and are at root based on some manifestation of substantive rationality. But there is no reason why intrinsic

rationality cannot be applied to these approaches as well. Satisficing should apply to the design of neural networks, for example, as well as to the design of a traditional differential equation-based model.

9.4   The enterprise of synthesis

The common task of physical science, philosophy, and psychology is to generate models of behavior (either physical or social) and to express these models as precisely as possible. Mathematics provides a language in which to couch these models. This language is essential to physical science and is at least convenient, to varying degrees, for philosophy and psychology.

A recent play by Michael Frayn, entitled *Copenhagen* (Frayn, 1998), is a hypothetical dialog between Bohr and Heisenberg, who were close collaborators in the first half of the nineteenth century. During World War II, Heisenberg, a German, paid a visit to Bohr, a Dane, in occupied Copenhagen. The content of their discussions has been a topic of great curiosity and speculation, and Frayn's play dramatizes the possibilities. It is in an interesting play, but for our purposes, only the following exchange from Act II is relevant:

Heisenberg: What something means is what it means in mathematics.
Bohr: You think that so long as the mathematics works out, the sense doesn't matter.
Heisenberg: Mathematics *is* sense. That's what sense is!
Bohr: But in the end, in the end, remember, we have to be able to explain it all to Margrethe!

(Margrethe is Bohr's wife, a non-scientific, non-mathematically proficient, but very articulate, lay person.) Bohr recognizes that scientific analysis is not complete until the written mathematical language of science has been translated into spoken language (conversation) and communicated to others. The essence of the above hypothetical conversation is captured by Wigner regarding scientific analysis: "The simplicities of natural laws arise through the complexities of the languages we use for their expression" (Wigner, 1960).

The concept dual to analysis is synthesis. Just as analysis is the primary business of science, synthesis is the primary business of engineering. Whereas science seeks to explain natural reality, engineering seeks to create an artificial – that is, man-made – reality. Much of engineering effort during the last century has concentrated on the design of entities that do relatively simple things in very limited contexts. With the explosion in computing capability and accessibility, however, the desire is growing to create entities that do more complicated things in larger contexts, such as group settings. Engineering and computer science literature is increasingly devoted to discussions about "intelligent" entities that can imitate some of the activities that humans might perform.

Humans govern their behavior by their concepts of rationality. But the question is, what concept or concepts do they use? There is no single answer. As with any scientific endeavor, many plausible theories may be proposed, and they may all be effective in explaining and predicting various behaviors, but there is no way to know for sure which, if any, of such analysis models is correct. Indeed, scientists, psychologists, and philosophers do not pretend to know the states of nature and mankind, they only seek theories that are consistent with observed behavior.

But the engineer must do more than hypothesize. It is not enough simply to concoct a plausible story line to explain behavior. *The artificial entity that is created must actually "live" the story line and perform the functions that are dictated by the model.* Consider another, hypothetical conversation between the protagonists of Frayn's play, this time reversing the roles of the characters.

Bohr: What something means is what it means in conversation.
Heisenberg: You think that so long as the conversation works out, the sense doesn't matter.
Bohr: Conversation *is* sense. That's what sense is!
Heisenberg: But in the end, in the end, remember, we have to be able to explain it all to the machine!

Explanations to a machine cannot be done with the spoken language of human conversation. They must be done with a language that is "understandable" to a computer. Mathematics is such a language, and when dealing with individual decision makers, the mathematics of optimization provides the ideal solution. But when dealing with social situations, the notion of what is "best" may not be well-defined, and we may need to consider "good enough," not merely as a substitute for the best, but as the ideal. Satisficing provides a new mathematics of amelioration. With it, order can emerge in a society through the natural local interactions that occur between agents who share common interests and who are willing to give deference to each other. Rather than depending upon the non-cooperative equilibria defined by individual rationality, satisficing may lead to the more socially realistic and valuable equilibria of shared interests and acceptable compromises.

Optimization is a very sophisticated notion of decision making that requires extrinsic total orderings, while satisficing is rather primitive notion of decision making that requires only intrinsic binary orderings. But this simplicity may be an important key. Perhaps, when considering the synthesis of multiple decision-making entities, we should form a dual to Wigner's sentiment: "The complexities of artificial entities arise through the simplicities of the components we use for their creation." It is the most fundamental component of all that this book addresses: the model of rationality that undergirds and overarches the design of artificial decision-making systems.

Appendix A: Bounded rationality

Bounded rationality deals with the problem of what to do when it is not possible or expedient to obtain a substantively rational solution owing to informational or computational limitations. Sims (1980) has characterized the situation as a research "wilderness." There are several different notions of bounded rationality, but they tend to follow two distinct streams. The first stream deals with bounds on the resources, such as computational power or time, and the second deals with bounds on the information available to obtain a decision.

Resource-bounded problems deal with the problem of making decisions under severe computational or time constraints. Consideration of such problems has led to the development of procedures that fall under a category termed **constrained optimization**. Constraints, in this context, do not refer to functional restrictions on the behavior of the decision maker, such as energy limitations or performance requirements. Rather, the constraints of interest here are limitations in the ability of the decision maker to process information and arrive at a solution. Within this subclass of problems, there are multiple approaches.

One way to deal with computational constraints is to modify the utility structure to maximize the comprehensive value of computation by accounting for time/computational cost as well as the intrinsic utility of the problem (Russell and Wefald, 1991; Kraus, 1996; Sandholm and Lesser, 1997). Another approach is to invoke a stopping rule, such that searching for an optimum solution proceeds until the cost of continuing to do so exceeds the expected benefits (Anderson and Milson, 1989; Hansen and Zilberstein, 1996). Although these approaches are attractive upon first glance, it can be difficult to balance the computational penalty with performance requirements or to evaluate the costs and benefits of continuing to search. Imposing constraints may offer more realistic expressions of the exigencies under which decisions must be made, but the results are nevertheless firmly couched in exactly the same premises as is unbounded (i.e., substantive) rationality. If the results are good enough, then they are good enough only because it is tacitly assumed that the best solution is always good enough. Essentially all that has changed is the criterion for being best.

Another view of bounded rationality is expressed by Kreps, who suggests that an apt definition for this concept is to be "intendedly [individually] rational, but limitedly so" (Kreps, 1990). This view admits the relaxation of the premise of total ordering but retains the premise of individual rationality. It is acknowledged, under this view, that informational limitations make it impossible to specify a complete ordering of premises (but, if one could, then one would certainly optimize). Essentially, the idea is to acquire a better understanding of the environment in which the agent is operating by learning from the past. Bicchieri (1993) recognizes that the individualistic, or micro, paradigm provides an inadequate explanation for group, or macro, phenomena, and argues that individual rationality must be supplemented by models of learning and by an evolutionary account of the emergent social order. A number of learning mechanisms have been suggested, including the methods of artificial intelligence such as neural computing, genetic algorithms, simulated anealing, and pattern recognition, with the expectation that such boundedly rational agents will learn to behave as if they were operating under conventional substantive rationality and at least achieve an approximately optimal solution (Sargent, 1993). Machine learning researchers have had some success in uncovering such processes (Bowling, 2000; Kalai and Lehrer, 1993; Fudenberg and Levine, 1993; Hu and Wellman, 1998).

A particular instantiation of this version of bounded rationality is found with **evolutionary game theory** (Maynard-Smith, 1982). Under this theory, "the game is played over and over again by biologically or socially conditioned players who are randomly drawn from large populations . . . and one assumes that some evolutionary selection process operates over time on the population distribution of behaviors" (Weibull, 1995). The evolutionary process requires two elements: a selection mechanism to establish preferences with respect to the varieties of available strategies, and a mutation mechanism to induce variety. The selection mechanism requires the imposition of a fitness ordering to govern the growth rates of the populations who employ the various strategies. Mutations are alternative strategies that randomly invade a population of decision makers; a strategy is evolutionarily stable if it is resistant (in terms of growth rates) to small invasions. The motivation for designing an evolutionary game is the hope that it will converge to an evolutionarily stable strategy that has desired characteristics (such as cooperative behavior). However, whereas the behavior of a von Neumann–Morgenstern game is expectant, in that the equilibria can be determined before the game is played, the behavior of an evolutionary game is temporally emergent and cannot be anticipated beforehand.

As with conventional von Neumann–Morgenstern theory, evolutionary game theory is built on the premises of rational choice. Bounding exists due to the relaxation (but not the relinquishment) of the total-ordering premise, and the individual rationality premise remains intact. Evolutionary game theory is a research area of growing importance. Although attention in this book is focused primarily on single-play decision problems, there is no conceptual reason why the development herein (which relaxes the premise of individual rationality) could not be extended to the multi-play case. Such an extension, however, is a topic for future research.

There are many other ways to form hybrid decision logics that blend the extremes of substantive and procedural rationality. *Ad hoc* juxtapositions of maximizing expectations and applying heuristics, however, are, at root, still heuristic, and the capacity for self-policing is compromised and may be destroyed.[1]

---

[1] For example, when designing a controller for a nonlinear dynamical system, a common practice is to linearize the system (a heuristic) and then design an optimal controller for the linearized system. There is no way to assure that the resulting design will function properly when applied to the nonlinear system.

Appendix B: Game theory basics

In this appendix we summarize the basic concepts of von Neumann–Morgenstern game theory that are relevant to our treatment of that topic in this book.

Definition B.1
An **agent**, or **player**, X, is a decision maker; that is, an entity that is capable of autonomously choosing from among a set of options. □

Definition B.2
A **decision problem** for X is a triple, (U, π, C), consisting of an option space, U (also called the action space), to be applied by the players, a set of outcomes or consequences, C, to be realized by the players, and a mediation mechanism, or mapping function, $\pi: U \to C$, that relates choices and outcomes. □

Definition B.3
A **joint decision problem**, or **game** for a family of agents, $\{X_1, \ldots, X_N\}$, where $N \geq 2$, is a triple $(\mathbf{U}, \boldsymbol{\pi}, \mathbf{C})$ where $\mathbf{U} = U_1 \times \cdots \times U_N$ is the joint action space with U_i being X_i's individual action space, $\mathbf{C} = C_1 \times \cdots \times C_N$ with C_i being X_i's individual consequence space, and $\boldsymbol{\pi} = (\pi_1, \ldots, \pi_N)$ where $\pi_i: \mathbf{U} \to C_i$ is a vector of mapping functions, one for each player.

An **option vector** is an ordered set of options, $\mathbf{u} = \{u_1, \ldots, u_N\}$, one for each player, where $u_i \in U_i$. □

Definition B.4
A **strategy**, s_i, for X_i is a rule for it to follow which tells it which option, $u_i \in U_i$, to choose. The **strategy space**, S_i, is the set of all strategies for X_i.

A **strategy vector** is is an ordered set of strategies, $\mathbf{s} = \{s_1, \ldots, s_N\}$, consisting of one strategy for each player. The **joint strategy space** is the product space $\mathcal{S} = S_1 \times \cdots \times S_N$. □

There is an important difference between an option and a strategy. Strictly speaking, a strategy is a complete set of instructions to tell the agent what option to take in every conceivable situation, even if it does not ever expect to reach that situation. For a multi-stage game (one that involves multiple moves by the players), a strategy for X_i is a sequence of options such that, at the kth stage, the kth element (which is an option) of the strategy is invoked. For single-stage games, a strategy is simply a rule to choose a single option from the option space. Another way to think about it is this: a strategy is a mental exercise that defines what should be done, and an action is the physical act of implementing the options.

The classical treatment of decision making and game theory was developed by von Neumann and Morgenstern (1944). This approach is based on utility theory. Utility theory requires each player to assign preferences to rank-order the various outcomes as a function of the possible strategy vectors.

Definition B.5

A **payoff function**, π_i, for X_i is a mapping from the joint strategy space to the real numbers. $\pi_i \colon \mathbf{S} \to \mathbb{R}$.   □

The payoff functions for all players are used to form an N-dimensional array such that each cell of the array contains an N-dimensional vector of payoff functions, one for each player, corresponding to each strategy vector. For example, consider a two-player single-stage game, each with two elementary options. Then the array becomes a 2×2 **payoff matrix** of the form given in Table 1.1.

Definition B.6

A **zero-sum** game is a game such that, for every strategy vector,

$$\sum_{i=1}^{N} \pi_i(\mathbf{s}) = 0.$$ □

Now that games have been formalized mathematically, we are in a position to discuss ways to solve them, that is, for each player to determine its strategy.

Definition B.7

An **equilibrium** is a strategy vector consisting of an acceptable strategy for each of the players.   □

Definition B.8

A **solution concept** is a rule that defines what it means to be acceptable, and hence defines an equilibrium.   □

There are several possible solution concepts leading to several kinds of equilibria. The four that are the most widely discussed are dominance, Nash equilibria, Pareto equilibria, and coordination equilibria.

Definition B.9

A strategy s_i^* is **dominant** for X_i if

$$\pi_i(s_1, \ldots, s_i^*, \ldots, s_N) > \pi_i(s_1, \ldots, s_i', \ldots, s_N)$$

for all $s_j \in S_j$, $j \neq i$, and all $s_i' \in U_i$ such that $s_i' \neq s_i^*$. Dominance, however, is a very strong property, and most interesting games do not have dominant strategies.   □

Definition B.10

A strategy vector $\mathbf{s}^* = \{s_1^*, \ldots, s_N^*\}$ is a **Nash equilibrium** if, were any single agent to change its strategy, it would reduce its payoff; that is, if

$$\pi_i(s_1^*, \ldots, s_i^*, \ldots, s_N^*) \geq \pi_i(s_1^*, \ldots, s_i', \ldots, s_N^*)$$

for all $s_i' \in S_i$ for $i = 1, \ldots, N$.   □

Definition B.11

A strategy vector $\mathbf{s}^* = \{s_1^*, \ldots, s_N^*\}$ is a **Pareto equilibrium** if no single agent, by changing its strategy, can increase its own payoff without lowering the payoff of at least one other agent; that is, if, for any i, $s_i' \neq s_i^*$ is such that

$$\pi_i(s_1^*, \ldots, s_i', \ldots, s_N^*) > \pi_i(s_1^*, \ldots, s_i^*, \ldots, s_N^*),$$

then

$$\pi_j(s_1^*, \ldots, s_i', \ldots, s_N^*) < \pi_j(s_1^*, \ldots, s_i^*, \ldots, s_N^*)$$

for some $j \neq i$.   □

Definition B.12

A strategy vector $\mathbf{s}^* = \{s_1^*, \ldots, s_N^*\}$ is a **coordination equilibrium** if no single agent would increase its payoff should *any one* agent alone act otherwise, either itself or someone else; that is, if $\mathbf{s}' = \{s_1^*, \ldots, s_j', \ldots, s_N^*\}$ for any j where where $s_j' \neq s_j^*$, then

$$\pi_i(\mathbf{s}') \leq \pi_i(\mathbf{s}^*)$$

for all $i = 1, \ldots, N$.   □

Appendix C: Probability theory basics

Probability theory developed as a means of analyzing the notion of chance and, as a mathematical discipline, it has developed in a rigorous manner based on a system of precise definitions and axioms. However, the syntax of probability theory exists independently of its use as a means of expressing and manipulating information and of quantifying the semantic notions of belief and likelihood regarding natural entities. In this book we employ the syntax of probability theory to quantify semantic notions that relate to the synthesis of artificial entities. In this appendix we present the basic notions of probability theory. However, since much of the terminology is motivated by the historical semantics, it will be necessary to supplement the standard treatment with terminology to render the definitions more relevant to our usage.

We begin by establishing the notation that is used in this book.

Definition C.1

A **set** is a collection of simple entities, called **elements**. If A is a set and the points $\omega_1, \omega_2, \ldots$ are its elements, we denote this relationship by the notation

$A = \{\omega_1, \omega_2, \ldots\}.$

If ω is an element of the set A, we denote this relationship by the notation $\omega \in A$, where $\in$ is the **element inclusion symbol**.

Often, we will specify a set by the properties of the elements. Suppose A comprises the set of all points ω such that $\omega \in S$ possesses property P. We express this relationship by the notation

$A = \{\omega: \text{Property } P \text{ holds}\}.$

For example, suppose A is the set of all ω such that $f(\omega) \geq g(\omega)$. We write this as

$A = \{\omega \in S: f(\omega) \geq g(\omega)\}.$

1. If ω is *not* an element of A, we express this relationship by the notation $\omega \notin A$.
2. If a set contains exactly one element ω, it is termed a **singleton** set, denoted $\{\omega\}$.

3. If A and B are sets and every element of B is also an element of A, we say that B is a **subset** of A or, equivalently, A is a **superset** of B, and denote this relationship by the notation

$$B \subset A \quad \text{or} \quad A \supset B,$$

where the symbols $\subset$ and $\supset$ are called **set inclusion symbols**.

4. If B is *not* a subset of A, we express this relationship by the notation $B \not\subset A$ or $A \not\supset B$.

5. If $\{\omega\}$ is a singleton set and ω is an element of A, we write $\{\omega\} \subset A$ to distinguish between the interpretations of ω as an element and $\{\omega\}$ as a set.

6. Let A and B be sets. The **union** of A and B is the set of elements that are members of either A or B (or both). We use the **union symbol** $\cup$ to denote this relationship; that is,

$$A \cup B = \{\omega: \omega \in A \text{ or } \omega \in B\}.$$

7. The **intersection** of A and B is the set of elements that are members of both A and B. We use the **intersection symbol** $\cap$ to denote this relationship; that is,

$$A \cap B = \{\omega: \omega \in A \text{ and } \omega \in B\}.$$

8. The **complement** A relative to B is the set of all elements that are in A but not in B. We use the **set minus symbol** $\setminus$ to denote this relationship, that is,

$$A \setminus B = \{\omega: \omega \in A \text{ and } \omega \notin B\}. \qquad \square$$

Definition C.2

The most fundamental concept of probability theory is that of an event. In the analysis context, an event corresponds to the state of the actual world.

1. An **elementary event** is the outcome of a simple experiment.

2. A **sample space**, Ω, is the set of all elementary events that may derive from an experiment.

3. A collection of elementary events is called an **event**. In particular, the entire sample space Ω is called the **sure event**, and the empty set $\emptyset$ is called the **null, or impossible, event**.

4. For any event A, its **complementary event**, denoted $\Omega \setminus A$, is also an event. $\qquad \square$

Using the same mathematical structure as conventional probability theory, we may form definitions suitable for a praxeological context. In this context, an event corresponds to the state of a possible world that would be realized if certain actions were taken. To distinguish between this and the standard probabilistic interpretation, we will refer to events in the praxeic context as options.

Definition C.3

An option corresponds to a possible action or collection of actions that a decision maker may implement.

1. An **elementary option** is a single action that that may be taken. We also refer to elementary options as **actions**.
2. An **option space**, U, is the set of all elementary options that may be taken by a decision maker.
3. A collection of elementary options is called an **option**. In particular, the entire action space U is called the **sure option**, and the empty set $\emptyset$ is called the **null, or impossible, option**.
4. For any option A, its **complementary option**, denoted $U \setminus A$, is also an option.   □

Definition C.4

A **Boolean algebra** of events (or options), denoted $\mathcal{F}$, is a class of events (options) in $\Omega(U)$ containing $\emptyset$ and closed under the operations of complementation and finite union. Thus, if $A_1 \in \mathcal{F}$ and $A_2 \in \mathcal{F}$, then $U \setminus A_1 \in \mathcal{F}$ and $A_1 \cup A_2 \in \mathcal{F}$.   □

Definition C.5

A σ-**field** of events (or options), also denoted $\mathcal{F}$, is a class of events (options) in $\Omega(U)$ containing $\emptyset$ and closed under the operations of complementation and countable union. Thus, if $A_i \in \mathcal{F}$, $i = 1, 2, \ldots$, then $U \setminus A_i \in \mathcal{F}$, $\bigcup_{i=1}^{\infty} A_i \in \mathcal{F}$.   □

There are a number of traditional semantic notions of probability in the analysis context, such as relative frequency, propensity, subjective belief, etc. These usages may be generally viewed as measures of truth support. Praxeic utility theory employs probability in two ways, both of which are different than any of the traditional usages. The first way is to use it as a measure of success support, which we term selectability. Thus, just as an event that is very likely to be true (i.e., to have occurred) will have a high probability value, an option that is very likely to succeed (i.e., to lead to the achievement of a goal) will have a high selectability value.

The second usage is that of a measure of the amount of resource that is consumed by adopting an option, which we term rejectability. Thus, an option that would consume a large fraction of the available resource would have a high rejectability value. We will assume that selectability and rejectability are semantic concepts that are defined independently of each other.

Definition C.6

A **probability measure**, denoted P_C, over a σ-field $\mathcal{F}$ of events in a sample space Ω is a function such that

$$P_C(\emptyset) = 0,$$
$$P_C(\Omega) = 1.$$

If A_i, $i = 1, 2, \ldots$ is a family of disjoint sets, then

$$P_C\left(\bigcup_{i=1}^{\infty} A_i\right) = \sum_{i=1}^{\infty} P_C(A_i).$$

A **selectability measure**, denoted P_S, is a probability measure over σ-field $\mathcal{F}$ of options. Likewise, a **rejectability measure**, denoted P_R, is also a probability measure over a σ-field $\mathcal{F}$ of options. $\quad\square$

Definition C.7

A **probability space** is a triple $(\Omega, \mathcal{F}, P_C)$. A **selectability space** is a triple $(U, \mathcal{F}, P_S)$. A **rejectability space** is a triple $(U, \mathcal{F}, P_R)$. $\quad\square$

Definition C.8

Let $(\Omega, \mathcal{F}, P_C)$ be a probability space such that Ω contains countably many elements, and let $\mathcal{F}$ be the power set of Ω (i.e., the set of all subsets of Ω). A **probability mass function**, p_C, is given by

$$p_C(\omega) = P_C(\{\omega\}),$$

where $\{\omega\}$ denotes the singleton set consisting of the elementary event ω for all $\omega \in \Omega$.

Let $(U, \mathcal{F}, P_S)$ be a selectability space such that U contains countably many elements, and let $\mathcal{F}$ be the power set of U. A **selectability mass function**, p_S, is given by

$$p_S(u) = P_S(\{u\})$$

for all elementary options $u \in U$.

Let $(U, \mathcal{F}, P_R)$ be a rejectability space such that U contains countably many elements, and let $\mathcal{F}$ be the power set of U. A **rejectability mass function**, p_R, is given by

$$p_R(u) = P_R(\{u\})$$

for all elementary options $u \in U$. $\quad\square$

Definition C.9

Let $(\Omega, \mathcal{F}, P_C)$ be a probability space such that Ω is a continuum of elementary events in $\mathbb{R}$, and let $\mathcal{F}$ be a σ-field over Ω. A **probability density function**, p_C, is the Radon–Nikodym derivative of P_C with respect to Lebesgue measure, and

$$P_C(A) = \int_A p_C(\omega)\,d\omega$$

for all $A \in \mathcal{F}$.

Let $(U, \mathcal{F}, P_S)$ be a selectability space such that U is a continuum of elementary options in $\mathbb{R}$, and let $\mathcal{F}$ be a σ-field over U. A **selectability density function**, p_S, is the Radon–Nikodym derivative of P_S with respect to Lebesgue measure, and

$$P_S(A) = \int_A p_S(u)du$$

for all $A \in \mathcal{F}$.

Let $(U, \mathcal{F}, P_S)$ be a rejectability space such that U is a continuum of elementary options in $\mathbb{R}$, and let $\mathcal{F}$ be a σ-field over U. A **rejectability density function**, p_R, is the Radon–Nikodym derivative of P_R with respect to Lebesgue measure, and

$$P_R(A) = \int_A p_R(u)du$$

for all $A \in \mathcal{F}$.   □

Definition C.10

Let U_1 and U_2 be option spaces. The **product option space** is the set of all ordered pairs of elememts of U_1 and U_2, respectively, and is denoted as

$$U_1 \times U_2 = \{(u, v): u \in U_1 \text{ and } v \in U_2\}.$$

If $A \subset U_1$ and $B \subset U_2$, the corresponding **product subset**, or **rectangle**, is the set of ordered pairs of elements of A and B, respectively, and is denoted

$$A \times B = \{(u, v): u \in A \text{ and } v \in B\}.$$ □

Definition C.11

Let A and B be events. The **conditional probability** of A given that B occurs is denoted $P(A|B)$. The symbol | is called the **conditioning symbol**.   □

Definition C.12

Let A and B be options.
1. The **conditional selectability** of A given that B is selected is denoted $P_{S|S}(A|B)$.
2. The **conditional selectability** of A given that B is rejected is denoted $P_{S|R}(A|B)$.
3. The **conditional rejectability** of A given that B is selected is denoted $P_{R|S}(A|B)$.
4. The **conditional rejectability** of A given that B is rejected is denoted $P_{R|R}(A|B)$.

                                                                               □

Appendix D: A logical basis for praxeic reasoning

The implication that probability theory has a praxeic role to play may appear to be an arbitrary and perhaps dubious appeal to a convenient analogy simply for purposes of posturing for credibility or enhancing intuition. However, the purpose of this appendix is to establish that the praxeic characterization arises from exactly the same kind of assumptions that underly the construction of probability theory for dealing with epistemic issues. We establish the claim that the mathematics of probability theory is the appropriate mechanism with which to characterize preference relationships between members of a multi-agent system. To facilitate this development, we will form praxeic analogies to the various epistemic concepts and present the epistemic concepts in parallel with the praxeic ones.

Probability is based on two-valued logic, i.e., Aristotelian logic. In the epistemic context, this means that an event is either true or false. For any event set A, we will say that A is true if any member of the set is true. If probability theory is to apply to praxeic considerations, it must be based on two-valued logic in that context as well. The praxeic Aristotelian analog to truth is instantiation – an action is either performed or it is not. For any set A of actions, we will say that A is instantiated if any member of the set is instantiated.

The goal of an epistemic inquiry is to ascertain truth. Analogously, the goal of a praxeic endeavour is to ascertain how to act. In this regard, there are two desiderata. The first is the achieving of the objective of the activity and the second is the conserving of resources. The first desideratum is, of course, related to the fundamental purpose of the endeavour, and the second desideratum exists because, in any practical situation, resources (money, fuel, time, exposure to hazard, etc.) must be expended in order to accomplish whatever tasks are appropriate in the pursuit of the first desideratum.

In the epistemic context, a basic notion to characterize an event in the quest for truth is *plausibility,* or the attractiveness of an option in terms of its veracity. There are two praxeic notions that are analogous to plausibility. The first is the notion of *suitability* to characterize the fitness of an action set in the interest of achieving the fundamental purpose of the endeavor. The second notion is that of *resistibility* to characterize the avoidance of an action set in the interest of conserving resources. The notions of suitability and resistibility must not be restatements of the same thing.

D.1   Desiderata for coherent evaluation

In the epistemic context we are interested in expressing degrees of plausibility regarding the truth of events under consideration, and in the praxeic context we are interested in expressing degrees of suitability and resistibility regarding the instantiation of the actions under consideration. Jaynes (2003) offers three basic desiderata that any theory of plausible reasoning ought to possess. The first desideratum is standard for all of formalized decision theory.

D-1 *Degrees of support are represented by real numbers.*

All degrees of support (in both the epistemic and praxeic contexts) must be defined in the context of the environment that pertains to the problem. Let e denote the environment (i.e., the state of nature). Since the plausibility (suitability, resistibility) of an event (option) may change as the environment changes, we will generally need to introduce additional notation to reflect this fact. We will use the notation $A \cdot e$ to denote the event A with respect to the environment e.[1] We say that $A \cdot e$ is true (instantiated) if any element of A is true (instantiated) and e is the environment. We say that $AB \cdot e$ is true (instantiated) if both $A \cdot e$ and $B \cdot e$ are true (instantiated) and we say that $A|B \cdot e$ is true (instantiated) if $A \cdot e$ is true (instantiated) given that $B \cdot e$ is true (instantiated).[2]

In accordance with Desideratum D-1, we must provide numerical values for preference orderings. Let us first consider the epistemic context and introduce the ordering $(\succeq_P, \cong_P)$ corresponding to ("is more plausible than," "is as plausible as"). We may then define a utility function $f_P(A \cdot e)$ corresponding to this ordering such that, if $A \cdot e \succeq_P B \cdot e$, then $f_P(A \cdot e) \geq f_P(B \cdot e)$.

In the praxeic context, we employ the orderings $(\succeq_S, \cong_S)$ and $(\succeq_R, \cong_R)$, meaning ("is more suitable than," "is as suitable as"), and ("is more resistible than," "is as resistible as"), respectively. Let $f_S(A \cdot e)$ and $f_R(A \cdot e)$ be utility functions denoting the respective numerical degrees of suitability and resistibility of A being instantiated under e.

We will assume that the functions f_P, f_S, and f_R all possess a continuity property, such that an infinitesimal change in the degrees of plausibility, suitability, and resistibility will generate only infinitesimal changes in the numerical values that the corresponding utility functions assume. When our discussion applies to all three utility functions, it will be convenient to suppress the subscript and let f represent a utility function for any of the contexts.

We now introduce the notion of *conditioning*, that is, the plausibility (suitability, resistibility) of one event (option), given that another event (option) is true (instantiated). Suppose that $B \cdot e$ is known to be true (instantiated). Then $f(A|B \cdot e)$ denotes the

---

[1] The symbol "$\cdot$" is the standard mathematical logic notation for the conjunction of two statements.
[2] We use the standard shorthand notation AB for $A \cap B$.

conditional plausibility (suitability, resistibility) of $A \cdot e$ being true (instantiated) given that $B \cdot e$ is true (instantiated).

To motivate the second desideratum, suppose e is changed to a new environment e' in such a way that $f(A \cdot e') > f(A \cdot e)$ but that $f(B|A \cdot e')$ is not changed. This should *never decrease* the plausibility (suitability, resistibility) of AB, that is, we require $f(AB \cdot e') \geq f(AB \cdot e)$. For example, if e changes in such a way as to make A more suitable but the suitability of B is not altered given that A is instantiated under either environment e or e', then the suitability of AB cannot decrease. Furthermore, if a change in e makes A more plausible (suitable, resistible) then it makes its complement less plausible (suitable, resistible); i.e., if $f(A \cdot e') > f(A \cdot e)$, then $f(A^c \cdot e') < f(A^c \cdot e)$, where A^c is the complement of A. We summarize this requirement with the following desideratum.

D-2 *Qualitative evaluations of support must agree with common sense.*

The third desideratum is a fundamental consistency argument.

D-3 *If a conclusion can be obtained in more than one way while considering exactly the same issues, then every possible way must lead to the same result.*

These three desiderata are sufficient to formulate specific quantitative rules of behavior that are suitable for governing both epistemic and praxeic activities.

D.2   Quantitative rules of behavior

We now seek a consistent rule for relating the plausibility (suitability, resistibility) of AB to the plausibilities (selectabilities, resistibilities) of A and B considered separately. We may evaluate the statement that A and B are both true (instantiated) under e by first considering $f(B \cdot e)$, the truth (instantiation) support for B under e, and then considering the statement that A is true (instantiated) given B under e, namely, $f(A|B \cdot e)$. Of course, since the intersection is commutative ($AB = BA$), we may reverse the roles of A and B without changing anything (this is in accordance with Desideratum D-3).

To elaborate, for a given environment, if both A and B are true (instantiated), then B must be true (instantiated). But if B is true (instantiated) then, for A also to be true (instantiated), it must be that A given B is true (instantiated). Furthermore, if B is false (not instantiated) then AB must also be false (not instantiated), regardless of whether or not A is true (instantiated). Thus, if we first consider the plausibility (suitability, resistibility) of B, then the plausibility (suitability, resistibility) of A will be relevant only if B is true (instantiated). Consequently, given the plausibilities (suitabilities, resistibilities) of both B and $A|B$, we do not require the plausibility (suitability, resistibility) of A alone in order to compute the plausibility (suitability, resistibility) of AB.

If we reverse the roles of A and B, then we see that the plausibility (suitability, resistibility) of A and $B|A$ is also sufficient to determine the plausibility (suitability, resistibility) of BA. The upshot of this development is that the plausibility (suitability, resistibility) of AB is a function of the plausibilities (suitabilities, resistibilities) of B and $A|B$ or, equivalently, of A and $B|A$. In other words,

$$f(AB \cdot e) = F[f(B \cdot e), f(A|B \cdot e)] = F[f(A \cdot e), f(B|A \cdot e)], \tag{D.1}$$

where F is some function to be determined.

Next, we observe that the consistency desideratum requires, when considering $f(ABC \cdot e)$, that if we consider BC first as a single event (option), application of (D.1) yields

$$f(ABC \cdot e) = F[f(BC \cdot e), f(A|BC \cdot e)]. \tag{D.2}$$

Next, we apply (D.1) to $f(BC \cdot e)$ to obtain

$$f(BC \cdot e) = F[f(C \cdot e), f(B|C \cdot e)]$$

which, when substituted into (D.2), yields

$$f(ABC \cdot e) = F\{F[f(C \cdot e), f(B|C \cdot e)], f(A|BC \cdot e)\}. \tag{D.3}$$

However, since set intersection is associative, we have $A(BC) = (AB)C$. Considering AB first, we also obtain

$$f(ABC \cdot e) = F[f(BC \cdot e), f(A|BC \cdot e)],$$

and repeated use of (D.1) also yields

$$f(ABC \cdot e) = F\{f(C \cdot e), F[f(B|C \cdot e), f(A|BC \cdot e)]\}. \tag{D.4}$$

The consistency desideratum requires that (D.3) and (D.4) must be the same. Thus, the function F must satisfy the following constraint, called the *associativity equation* (Jaynes, 2003; Aczél, 1966):

$$F[F(x, y), z] = F[x, F(y, z)]. \tag{D.5}$$

A further constraint on F is that it also must satisfy the common sense desideratum. To ensure this property, suppose e changes to e' such that $f(B \cdot e') > f(B \cdot e)$ but $f(A|B \cdot e') = f(A|B \cdot e)$. Then common sense insists that $f(AB \cdot e') \geq f(AB \cdot e)$. Also, we require that, if $f(B \cdot e') = f(B \cdot e)$ but $f(A|B \cdot e') > f(A|B \cdot e)$, then $f(AB \cdot e') \geq f(AB \cdot e)$. In other words, $F(x, y)$ must be non-decreasing in both arguments.

It remains to determine the structure of F that satisfies all of these constraints. By direct substitution, it is easily established that (D.5) is satisfied if

$$w[F(x, y)] = w(x) \cdot w(y) \tag{D.6}$$

for any function w. The following theorem establishes this as the general solution.

Theorem D.1

(Cox, 1946) *Suppose F is differentiable in both arguments, then (D.6) is the general solution to (D.5) for some positive, continuous, monotonic function w. Consequently, for any events (options) A and B,*

$$w[f(AB \cdot e)] = w[f(A|B \cdot e)] \cdot w[f(B \cdot e)], \tag{D.7}$$

which is called the product rule, *and*

$$w[f(A \cdot e)] + w[f(A^c \cdot e)] = 1,$$

which is called the sum rule. *Furthermore,*

$$w[f(U \cdot e)] = 1$$

and

$$w[f(A \cup B \cdot e)] = w[f(A \cdot e)] + w[f(B \cdot e)] - w[f(AB \cdot e)].$$

For a proof of this theorem see Cox (1946, 1961), Tribus (1969), or Jaynes (2003). Jaynes observes that (D.5) was actually first solved by Abel as early as 1826 in a different context (Abel, 1881). Also, Aczél has established the same result without the assumption of differentiability (Aczél, 1966).

D.3   Constructing probability (selectability, rejectability)

Lets us now impose the additional assumption that w is non-decreasing. Also, since w composed with f is a function of the events (options), we may, without loss of generality, define a function P over the events (options) as $P(A) = w[f(A \cdot e)]$ (since we now assume the environment is fixed, we may simplify notation by dropping e from the argument list). Then P possesses exactly the mathematical properties that are required to define probability over a Boolean algebra of events (options); namely,

P-1 *Non-negativity*: $0 \leq P(A)$

P-2 *Normalization*: If U is the entire space, then $P(U) = 1$

P-3 *Additivity*: If A and B are disjoint, then $P(A \cup B) = P(A) + P(B)$

for A, B in $\mathcal{B}$. These three properties are usually taken as axioms in standard expositions of probability theory *à la* Kolmogorov (1956). These axioms are then used to derive all of the well-known probabilistic concepts, such as conditioning and independence. However, we see that Kolmogorov's axioms themselves are actually the consequences of more descriptive desiderata. In particular, conditioning, which is expressed merely as a *definition* with conventional treatments, is actually a fundamental concept under the more constructive approach offered by Cox. That is, (D.7) becomes the product rule of probability theory:

$$P(AB) = P(A|B) \cdot P(B). \tag{D.8}$$

We conclude that the mathematical structure of probability theory is a valid characterization of uncertainty, whether the context pertains to the epistemic notion of characterizing the truth support of events or to the praxeic notions of characterizing the success support and resource consumption support of options. To emphasize the praxeic context, we will refer to P_S and P_R as selectability and rejectability functions, respectively, rather than probability functions. $P_S(A)$ is the degree of support for selecting A in the interest of achieving the goal of the decision maker, and $P_R(A)$ is the degree of support for rejecting A in the interest of conserving resources. Furthermore, we may view these two functions as marginals of an even more general function, which we term the *interdependence* function. This function, denoted P_{SR}, is defined over options in the product space $U \times U$ such that, for any options $A \subset U$ and $B \subset U$, the functions P_S and P_R are marginals of P_{SR}:

$$P_S(A) = P_{SR}(A \times U), \tag{D.9}$$
$$P_R(B) = P_{SR}(U \times B). \tag{D.10}$$

Let p_{SR} be the mass function associated with P_{SR} (that is, for any singleton pair $(u, v) \in U \times U$, $p_{SR}(u; v) = P_{SR}(\{u\} \times \{v\})$). We will term p_{SR} the *interdependence mass function*. Then p_S and p_R are the selectability and rejectability marginals of p_{SR}, that is, $p_S(u) = \sum_{v \in U} p_{SR}(u; v)$ and $p_R(v) = \sum_{u \in U} p_{SR}(u; v)$. In the general n-agent case, the action space U will be n-dimensional, and u and v will be n-dimensional vectors.

Bibliography

Abel, N. H. (1881). *Oeuvres Complètes de Niels Henrik Abel*, editors Sylow, L. and Lie, S. Christiania: de Groendahl and soen.

Aczél, J. (1966). *Lectures on Functional Equations and Their Applications*. Academic Press, New York.

Anderson, J. R. and Milson, R. (1989). Human memory: An adaptive perspective. *Psychological Review*, 98:703–19.

Arrow, K. J. (1950). A difficulty in the concept of social welfare. *Journal of Political Economy*, 58.

Arrow, K. J. (1951). *Social Choice and Individual Values*. John Wiley, New York. 2nd edn. 1963.

Arrow, K. J. (1986). Rationality of self and others. In Hogarth, R. M. and Reder, M. W., editors, *Rational Choice*. University of Chicago Press, Chicago.

Ash, R. (1965). *Information Theory*. John Wiley, New York.

Axelrod, R. (1984). *The Evolution of Cooperation*. Basic Books, New York.

Axelrod, R. (1997). *The Complexity of Cooperation*. Princeton University Press, Princeton, NJ.

Azoulay-Schwartz, R. and Kraus, S. (2002). Negotiation on data allocation in multi-agent environments. *Autonomous Agents and Multi-Agent Systems Journal*, 5:123–72.

Bacharach, M. (1976). *Economics and the Theory of Games*. Macmillan, London.

Ball, L. J., Maskill, L., and Ormerod, T. C. (1996). Satisficing in engineering design: Causes, consequences and implications for design support. In *Proceedings of the First International Symposium on Descriptive Models of Design*, pages 317–32, Istanbul.

Bazerman, M. (1983). A critical look at the rationality of negotiator judgement. *Behavorial Science*, 27:211–28.

Bazerman, M. H. and Neale, M. A. (1992). Negotiator rationality and negotiator cognition: The interactive roles of prescriptive and descriptive research. In Young, P. H., editor, *Negotiation Analysis*, pages 109–29. University of Michigan Press, Ann Arbor, MI.

Beeson, D. (1992). *Maupertuis: an Intellectual Biography*. The Alden Press, Oxford.

Bergson, A. (1938). A reformulation of certain aspects of welfare economics. *Quarterly Journal of Economics*, 52:310–34.

Bestougeff, H. and Rudnianski, M. (1998). Games of deterrence and satisficing models applied to business process modeling. In *Proceedings of the 1998 AAAI Symposium*, pages 8–14. March 23–25, Stanford California. Technical Report SS-98-05.

Bicchieri, C. (1993). *Rationality and Coordination*. Cambridge University Press, Cambridge.

Binmore, K. G. (1998). *Game Theory and the Social Contract: Just Playing*. MIT Press, Cambridge, MA.

Bowling, M. (2000). Convergence problems of general-sum multiagent reinforcement learning. In *Proceedings of the Seventeenth International Conference on Machine Learning*.

Brams, S. J. (1990). *Negotiation Games: Applying Game Theory to Bargaining and Arbitration.* Routledge, New York.

Browning, R. (1855). Andrea del sarto. In *Men and Women.* Chapman and Hall, London.

Castellan, N. J., editor (1993). *Individual and Group Decision Making.* Lawrence Erlbaum Associates, Hillsdale, NJ.

Casti, J. L. (1997). *Would-Be Worlds: How Simulation is Changing the Frontiers of Science.* Wiley, New York.

Cohen, P. and Levesque, H. (1990). Intention is choice with commitment. *Artificial Intelligence,* 42:262–310.

Cohen, P., Morgan, J., and Pollack, M. E., editors (1990). *Intentions in Communication.* MIT Press, Cambridge, MA.

Cohen, P. R. (1990a). Architectures and strategies for reasoning under uncertainty. In Shafer, G. and Pearl, J., editors, *Readings in Uncertain Reasoning,* pages 167–176. Morgan Kaufmann, San Mateo, CA.

Cohen, P. R. (1990b). The control of reasoning under uncertainty: A discussion of some programs. In Shafer, G. and Pearl, J., editors, *Readings in Uncertain Reasoning,* pages 177–197. Morgan Kaufmann, San Mateo, CA.

Conry, S. E., Kuwabara, K., Lesser, V. R., and Meyer, R. A. (1991). Multistage negotiation for distributed constraint satisfaction. *IEEE Transactions on Systems, Man, and Cybernetics,* 21(6):1462–77.

Cooper, R., DeJong, D. V., Forsythe, R., and Ross, T. W. (1996). Cooperation without reputation: experimental evidence from prisoner's dilemma games. *Games and Economic Behavior,* 12(1):187–218.

Cover, T. M. and Thomas, J. A. (1991). *Elements of Information Theory.* John Wiley, New York.

Cox, R. T. (1946). Probability, frequency, and reasonable expectation. *American Journal of Physics,* 14:1–13.

Cox, R. T. (1961). *The Algebra of Probable Inference.* Johns Hopkins University Press, Baltimore.

Cyert, R. M. and March, J. G. (1992). *A Behavorial Theory of the Firm.* Blackwell Publishers, Cambridge, MA, second edition.

de Maupertuis, P.-L. M. (1740). Loi du repos des corps. In *Mémoires de l'Académie des sciences.*

DeGroot, M. (1970). *Optimal Statistical Decisions.* McGraw-Hill, New York.

Durfee, E. H. and Lesser, V. R. (1989). Negotiating task decomposition and allocation using partial global planning. In Gasser, L. and Huhns, M. N., editors, *Distributed Artificial Intelligence,* volume II, pages 229–44. Pitman/Morgan Kaufmann, London.

Durfee, E. H., Lesser, V. R., and Corkill, D. D. (1989). Cooperative distributed problem solving. In Barr, A., Cohen, P. R., and Feigenbaum, E. A., editors, *The Handbook of Artificial Intelligence,* volume 4, pages 83–148. Morgan Kauffman.

Edwards, W. (1990). Unfinished tasks: A research agenda for behavioral decision theory. In Hogarth, R. M., editor, *Insights in Decision Making.* University Chicago Press, Chicago.

Ehtamo, H., Verkama, M., and Hämäläinen, R. P. (1999). How to select fair improving directions in a negotiation model over continuous issues. *IEEE Transactions on Systems, Man, and Cybernetics,* 29(1):26–33.

Eisen, M. (1969). *Introduction to Mathematical Probability Theory.* Prentice-Hall, Englewood Cliffs, NJ.

Ephrati, E. and Rosenschein, J. S. (1996). Deriving consensus in multiagent systems. *Artificial Intelligence,* 87(1–2):21–74.

Erlandson, R. F. (1981). The satisficing process: A new look. *IEEE Transactions on Systems, Man, and Cybernetics,* 11(11):740–52.

Ferguson, T. S. (1967). *Mathematical Statistics*. Academic Press, New York.

Franklin, B. (1772/1987). *Writings*. The Library of America, New York. Original letter written Sept. 19, 1772.

Frayn, M. (1998). *Copenhagen*. Methuen Drama, London.

Fudenberg, D. and Levine, D. K. (1993). Steady state learning and Nash equilibrium. *Econometrica*, 61(3):547–73.

Gale, J., Bimore, K., and Samuelson, L. (1995). Learning to be imperfect: the ultimatum game. *Games and Economic Behavior*, 8:56–90.

Gigerenzer, G. and Goldstein, D. G. (1996). Reasoning the fast and frugal way: Models of bounded rationality. *Psychological Review*, 103(4):650–69.

Gigerenzer, G. and Todd, P. M. (1999). *Simple Heuristics that Make Us Smart*. Oxford University Press, New York.

Glass, A. and Grosz, B. (2000). Socially conscious decision-making. In *Proceedings of Agents 2000 Conference*, pages 217–24, Barcelona, Spain.

Gmytrasiewicz, P. J., Durfee, E. H., and Wehe, D. K. (1991). The utility of communication in coordinating intelligent agents. In *Proceedings of the Ninth National Conference on Artificial Intelligence*, pages 116–72.

Goldstein, W. M. and Hogarth, R. M. (1997). *Research on Judgment and Decision Making*. Cambridge University Press, Cambridge.

Goodrich, M. A. (1996). On a Theory of Satisficing Control. PhD dissertation, Brigham Young University.

Goodrich, M. A., Stirling, W. C., and Boer, E. R. (2000). Satisficing revisited. *Minds and Machines*, 10:79–109.

Goodrich, M. A., Stirling, W. C., and Frost, R. L. (1996). A satisficing approach to intelligent control of nonlinear systems. In *Proceedings of the IEEE Symposium on Intelligent Control*, pages 248–52, Dearborn, Michigan.

Goodrich, M. A., Stirling, W. C., and Frost, R. L. (1998). A theory of satisficing decision and control. *IEEE Transactions on Systems, Man, and Cybernetics*, 28(6):763–79.

Goodrich, M. A., Stirling, W. C., and Frost, R. L. (1999). Model predictive satisficing fuzzy logic control. *IEEE Transaction on Fuzzy Systems*, 7(3):319–32.

Hansen, E. A. and Zilberstein, S. (1996). Monitoring the progress of anytime problem-solving. In *Proceedings of the 13th National Converence on Artificial Intelligence*, pages 1229–34. Portland, Oregon.

Hardin, G. (1968). The tragedy of the commons. *Science*, 162:1243–8.

Harsanyi, J. (1955). Cardinal welfare, individualistic ethics, and interpersonal comparisons of utility. *Journal of Political Economy*, 63:315.

Harsanyi, J. (1977). *Rational Behavior and Bargaining Equilibrium in Games and Social Situations*. Cambridge University Press, Cambridge.

Hart, S. and Mas-Colell, A. (1994). *Cooperation: Game-Theoretic Approaches*. Springer-Verlag, Berlin.

Hicks, J. R. (1939). Foundations of welfare economics. *Economic Journal*.

Ho, Y. C. (1994). Heuristics, rules of thumb, and the 80/20 proposition. *IEEE Transactions on Automatic Control*, 39(5):1025–7.

Ho, Y. C. (1997). On the numerical solutions of stochastic optimization problem. *IEEE Transactions on Automatic Control*, 42(5):727–9.

Hogarth, R. M. and Reder, M. W. (1986a). Perspectives from economics and psychology. In Hogarth, R. M. and Reder, M. W., editors, *Rational Choice*. University of Chicago Press, Chicago.

Hogarth, R. M. and Reder, M. W., editors (1986b). *Rational Choice*. University of Chicago Press, Chicago.

Hogg, L. and Jennings, N. R. (1999). Variable sociability in agent-based decision making. In Parsons, S. and Wooldridge, M. J., editors, *Workshop on Decision Theoretic and Game Theoretic Agents*, pages 29–42. University College, London, United Kingdom, 5 July.

Horvitz, E. and Zilberstein, S. (2001). Computational tradeoffs under bounded resources (editorial). *Artificial Intelligence*, 126:1–4.

Hu, J. and Wellman, M. P. (1998). Multiagent reinforcement learning: Theoretical framework and an algorithm. In Shavlik, J., editor, *Proceedings of the Fifteenth International Conference on Machine Learning*, pages 242–50.

James, W. (1956). *The Will to Believe and Other Essays*. Dover, New York.

Jaynes, E. T. (2003). *Probability Theory: The Logic of Science*. Cambridge University Press, Cambridge.

Johnson-Laird, P. N. (1988). *The Computer and the Mind: An Introduction to Cognitive Science*. Harvard University Press, Cambridge, MA.

Kahenman, D., Knetsch, J. L., and Thaler, R. H. (1986). Fairness and the assumptions of economics. In Hogarth, R. M. and Reder, M. W., editor, *Rational Choice*. University of Chicago Press, Chicago.

Kalai, E. and Lehrer, E. (1993). Rational learning leads to Nash equilibrium. *Econometrica*, 61(5):1019–45.

Kaldor, N. (1939). Welfare propositions of economic and interpersonal comparisons of utility. *Economic Journal*.

Kaufman, B. E. (1990). A new theory of satisficing. *The Journal of Behavorial Economics*, 19(1):35–51.

Kirk, D. (1970). *Optimal Control Theory*. Prentice-Hall, Englewood Cliffs, NJ.

Kohn, A. (1992). *No Contest: The Case Against Competition*. Houghton Mifflin, Boston. Revised edition.

Kolmogorov, A. N. (1956). *Foundations of the Theory of Probability*. Chelsea Publishing Company, New York, second edition. German original 1933.

Kotarbiński, T. (1965). *Praxiology: An Introduction to the Sciences of Efficient Action*. Pergamon Press, Oxford. Translated by Olgierd Wojtasiewicz.

Kraus, S. (1996). Beliefs, time and incomplete information in multiple encounter negotiations among autonomous agents. *Annals of Mathematics and Artificial Intelligence*, 20(1–4):111–59.

Kraus, S. and Lehmann, D. (1999). Knowledge, belief and time. *Theoretical Computer Science*, 58:155–74.

Kraus, S., Sycara, K., and Evenchik, A. (1998). Reaching agreements through argumentation: a logical model and implementation. *Artificial Intelligence Journal*, 104(1–2):1–69.

Kraus, S. and Wilkenfeld, J. (1991). Negotiations over time in a mutiagent environment. In *Proceedings of IJCAI-91*, pages 56–61.

Kraus, S. and Wilkenfeld, J. (1993). A strategic negotiations model with applications to an international crisis. *IEEE Transactions on Systems, Man, and Cybernetics*, 32(1):313–23.

Kraus, S., Wilkenfeld, J., and Zlotkin, G. (1995). Multiagent negotiation under time constraints. *Artificial Intelligence*, 75(2):297–345.

Kreps, D. M. (1990). *Game Theory and Economic Modelling*. Clarendon Press, Oxford.

Krovi, R., Graesser, A. C., and Pracht, W. E. (1999). Agent behaviors in virtual negotiation environments. *IEEE Transactions on Systems, Man, and Cybernetics*, 29(1):15–25.

Levi, I. (1967). *Gambling with Truth*. MIT Press, Cambridge, MA.

Levi, I. (1980). *The Enterprise of Knowledge*. MIT Press, Cambridge, MA.

Levi, I. (1982a). Conflict and social agency. *Journal of Philosophy*, 79(5):231–46.

Levi, I. (1982b). Ignorance, probability, and rational choice. *Synthese*, 53:387–417.

Levi, I. (1984). *Decisions and Revisions*. Cambridge University Press, Cambridge.

Levi, I. (1985). Imprecision and indeterminacy in probability judgement. *Philosophy of Science*, 52(3):390–409.

Levi, I. (1986). *Hard Choices*. Cambridge University Press, Cambridge.

Levi, I. (1991). *The Fixation of Belief and Its Undoing*. Cambridge University Press, Cambridge.

Levi, I. (1997). *The Covenant of Reason*. Cambridge University Press, Cambridge.

Lewis, D. K. (1969). *Convention*. Harvard University Press, Cambridge, MA.

Lewis, F. L. (1986). *Optimal Control*. Wiley-Interscience, New York.

Lewis, M. and Sycara, K. (1993). Reaching informed agreement in multi-specialist cooperation. *Group Decision and Negotiation*, 2(3):279–300.

Lilly, G. (1994). Bounded rationality. *Journal of Economic Dynamics and Control*, 18:205–30.

Luce, R. D. and Raiffa, H. (1957). *Games and Decisions*. John Wiley, New York.

Malone, T. W. (1987). Modeling coordination in organizations and markets. *Management Science*, 33(10):1317–32.

Mansbridge, J. J., editor (1990a). *Beyond Self-Interest*. University of Chicago Press, Chicago.

Mansbridge, J. J. (1990b). The relation of altruism and self-interest. In Mansbridge, J. J., editor, *Beyond Self-Interest*, Chapter 8. University of Chicago Press, Chicago.

Margolis, H. (1990). Dual utilities and rational choice. In Mansbridge, J. J., editor, *Beyond Self-Interest*, chapter 15. University of Chicago Press, Chicago.

Matsuda, T. (1979a). Algebraic properties of satisficing decision criterion. *Information Sciences*, 17:221–37.

Matsuda, T. (1979b). Characterization of satisficing decision criterion. *Information Sciences*, 17:131–51.

Maynard-Smith, J. (1982). *Evolution and the Theory of Games*. Cambridge University Press, Cambridge.

Moon, T. K., Budge, S. E., Stirling, W. C., and Thompson, J. B. (1994). Epistemic decision theory applied to multiple-target tracking. *IEEE Transactions on Systems, Man, and Cybernetics*, SMC-24(2):234–45.

Moon, T. K., Frost, R. L., and Stirling, W. C. (1996). An epistemic utility approach to coordination in the prisoner's dilemma. *BioSystems*, 37(1–2):167–76.

Moore, J. C. and Rao, H. R. (1994). Multi-agent resource allocation: An incomplete information perspective. *IEEE Transactions on Systems, Man, and Cybernetics*, SMC-14(8):1208–19.

Morgan, M. G. and Henrion, M. (1990). *Uncertainty*. Cambridge University Press, Cambridge.

Morrell, D. R. (1988). A Theory of Set-valued Estimation. PhD thesis, Brigham Young University.

Morrell, D. R. and Stirling, W. C. (1991). Set-valued filtering and smoothing. *IEEE Trans. Systems, Man, Cybernet*, 21(1):184–93.

Mullen, T. and Wellman, M. (1995). A simple computational market for network information services. In *Proceedings of the First International Conference on Multiagent Systems*, pages 283–9.

Nanson, E. J. (1882). Methods of elections. *Transactions and Proceedings of the Royal Society of Victoria*, 18.

Nash, J. F. (1950). The bargaining problem. *Econometrica*, 18:155–62.

Nash, J. F. (1951). Non-cooperataive games. *Annals of Mathematics*, 54:289–95.

Neveu, J. (1965). *Mathematical Foundations of the Calculus of Probability*. Holden Day, San Francisco.

Newell, A. (1990). *Unified Theories of Cognition*. Harvard University Press, Cambridge, MA.

Nowak, M. A., Page, K. M., and Sigmund, K. (2000). Fairness versus reason in the ultimatum game. *Science*, 289:1773–5.

Oaksford, M. and Chater, N. (1994). A rational analysis of the selection task as optimal data selection. *Psychological Review*, 101:608–31.

Osborne, M. J. and Rubinstein, A. (1994). *A Course in Game Theory*. MIT Press, Cambridge, MA.

Pearce, D. W. (1983). *Cost-Benefit Analysis*. Macmillan, London, second edition.

Pearl, J. (1988). *Probabilistic Reasoning in Intelligent Systems*. Morgan Kaufmann, San Mateo, CA.

Peirce, C. S. (1877). The fixation of belief. *Popular Science Monthly*, 12.

Polanyi, M. (1962). *Personal Knowledge*. University Chicago Press, Chicago.

Polya, G. (1954). *Induction and Analogy in Mathematics*. Princeton University Press, Princeton, NJ.

Popper, K. R. (1963). *Conjectures and Refutations: The Growth of Scientific Knowledge*. Harper and Row, New York.

Raiffa, H. (1968). *Decision Analysis*. Addison-Wesley, Reading, MA.

Raiffa, H. (1982). *The Art and Science of Negotiation*. Harvard University Press, Cambridge, MA.

Rapoport, A. (1970). *N-Person Game Theory*. The University of Michigan Press, Ann Arbor, MI.

Rapoport, A. and Orwant, C. (1962). Experimental games: a review. *Behavorial Science*, 7:1–36.

Rasmusen, E. (1989). *Games and Information*. Basil Blackwell, Oxford.

Rosenschein, J. and Zlotkin, G. (1994). *Rules of Encounter*. MIT Press, Cambridge, MA.

Ross, S. (2002). *A First Course in Probability*. Prentice-Hall, Upper Saddle River, N.J., 6th edition.

Roth, A. E. (1995). Bargaining experiments. In Kagel, J. H. and Roth, A. E., editors, *Handbook of Experimental Economics*, Chapter 4. Princeton University Press, Princeton, NJ.

Rubenstein, A. (1982). Perfect equilibrium in a bargaining model. *Econometrica*, 50(1):97–109.

Rubinstein, A. (1998). *Modeling Bounded Rationality*. MIT Press, Cambridge, MA.

Russell, S. and Wefald, E. (1991). Principles of metareasoning. *Artificial Intelligence*, 49:361–95.

Samuelson, P. A. (1948). *Foundations of Economic Analysis*. Harvard University Press, Cambridge, MA.

Sandholm, T. W. (1999). Distributed rational decision making. In Weiss, G., editor, *Multiagent Systems*, chapter 5, pages 201–258. MIT Press, Cambridge, MA.

Sandholm, T. and Lesser, V. (1997). Coalitions among computationally bounded agents. *Artificial Intelligence*, 94(1):99–137.

Sargent, T. J. (1993). *Bounded Rationality in Macroeconomics*. Oxford University Press, Oxford.

Schelling, T. C. (1960). *The Strategy of Conflict*. Oxford University Press, Oxford.

Scitovsky, T. (1941). A note on welfare propositions in economics. *Review of Economic Studies*.

Sen, A. K. (1979). *Collective Choice and Social Welfare*. North-Holland, Amsterdam.

Sen, A. K. (1990). Rational fools: A critique of the behavorial foundations of economic theory. In Mansbridge, J. J., editor, *Beyond Self-Interest*, Chapter 2. University of Chicago Press, Chicago.

Sen, S., editor (1998). *Satisficing Models*. AAAI Press, San Mateo, CA.

Sen, S. and Durfee, E. H. (1994). The role of commitment in cooperative negotiation. *International Journal of Intelligent and Cooperative Information Systems*, 3(1):67–81.

Shafer, G. (1976). *A Mathematical Theory of Evidence*. Princeton University Press, Princeton, NJ.

Shafer, G. and Pearl, J., editors (1990). *Readings in Uncertain Reasoning*. Morgan Kaufmann, San Mateo, CA.

Shapley, L. S. (1953). A value for n-person games. In Kuhn, H. W. and Tucker, A. W., editors, *Contributions to the Theory of Games*. Princeton University Press, Princeton, NJ.

Shoham, Y. (1993). Agent oriented programming. *Artificial Intelligence*, 60(1):51–92.

Shubik, M. (1982). *Game Theory in the Social Sciences*. MIT Press, Cambridge, MA.

Sigmund, K. (1995). *Games of Life: Explorations in Ecology, Evolution and Behavior*. Penguin, London.

Sigmund, K., Fehr, E., and Nowak, M. A. (2002). The economics of fair play. *Scientific American*, 286:83–7.

Simon, H. A. (1955). A behavioral model of rational choice. *Quarterly Journal of Economics*, 59:99–118.

Simon, H. A. (1956). Rational choice and the structure of the environment. *Psychological Review*, 63(2):129–38.

Simon, H. A. (1982a). *Models of Bounded Rationality: Behavorial Economics and Business Organizations*. MIT Press, Cambridge, MA.

Simon, H. A. (1982b). *Models of Bounded Rationality: Economic Analysis and Public Policy*. MIT Press, Cambridge, MA.

Simon, H. A. (1986). Rationality in psychology and economics. In Hogarth, R. M. and Reder, M. W., editors, *Rational Choice*. University Chicago Press, Chicago.

Simon, H. A. (1996). *The Sciences of the Artificial*. MIT Press, Cambridge, MA, third edition.

Simon, H. A. (1997). *Models of Bounded Rationality: Empirically Grounded Economic Reason*. MIT Press, Cambridge, MA.

Sims, C. (1980). Macroeconomics and reality. *Econometrica*, 69:1–48.

Skyrms, B. (1990). *The Dynamics of Rational Deliberation*. Harvard University Press, Cambridge, MA.

Slote, M. (1989). *Beyond Optimizing*. Harvard University Press, Cambridge, MA.

Sober, E. and Wilson, D. S. (1998). *Unto Others: The Evolution and Psychology of Unselfish Behavior*. Harvard University Press, Cambridge, MA.

Stirling, W. (1994). Multi-agent coordinated decision-making using epistemic utility theory. In Castelfranchi, E. and Werner, E., editors, *Artificial Social Systems*, pages 164–83. Springer-Verlag, Berlin.

Stirling, W. C. (1991). A model for multiple agent decision making. In *Proceedings of the 1991 IEEE Conference on Systems, Man, and Cybernetics*, pages 2073–8.

Stirling, W. C. (1992). Coordinated intelligent control via epistemic utility theory. In *Proceedings of the 1992 International Conference on Systems, Man, and Cybernetics*, pages 37–42.

Stirling, W. C. (1993). Coordinated intelligent control via epistemic utility theory. *IEEE Control Systems Magazine*, 13(5):21–9.

Stirling, W. C. and Frost, R. L. (1994). Making value-laden decisions under conflict. In *Proceedings of the 1994 IEEE Conference on Systems, Man, and Cybernetics*.

Stirling, W. C. and Frost, R. L. (2000). Intelligence with attitude. In *Proceedings of the Performance Metrics for Intelligent Systems Workshop*, Gaithersburg, MD. National Institute of Standards and Technology.

Stirling, W. C. and Goodrich, M. A. (1999a). Satisficing equilibria: A non-classical approach to games and decisions. In Parsons, S. and Wooldridge, M. J., editors, *Workshop on Decision Theoretic and Game Theoretic Agents*, pages 56–70, University College, London, United Kingdom, 5 July.

Stirling, W. C. and Goodrich, M. A. (1999b). Satisficing games. *Information Sciences*, 114:255–80.

Stirling, W. C. and Morrell, D. R. (1991). Convex bayes decision theory. *IEEE Transactions on Systems, Man, and Cybernetics*, 21(1):173–83.

Stirling, W. C., Goodrich, M. A., and Frost, R. L. (1996a). Procedurally rational decision-making and control. *IEEE Control Systems Magazine*, 16(5):66–75.

Stirling, W. C., Goodrich, M. A., and Frost, R. L. (1996b). Satisficing intelligent decisions using epistemic utility theory. In *Intelligent Systems: A Semiotic Perspective. Proceedings of the 1996*

International Multidisciplinary Conference Conference on Multi-Agent Systems, Volume II: Applied Semiotics, pages 268–73.

Stirling, W. C., Goodrich, M. A., and Frost, R. L. (1996c). Toward a theoretical foundation for multi-agent coordinated decisons. In *Proceedings of The Second International Conference on Multi-Agent Systems*, pages 345–52, Kyoto, Japan.

Stirling, W. C., Goodrich, M. A., and Packard, D. J. (2002). Satisficing equilibria: A non-classical approach to games and decisions. *Autonomous Agents and Multi-Agent Systems Journal*, 5:305–28.

Straffin, P. (1980). The prisoner's dilemma. *UMAP Journal*, 1:101–3.

Sycara, K. (1990). Persuasive argumentation in negotiation. *Theory and Decision*, 28:203–42.

Sycara, K. (1991). Problem restructuring in negotiation. *Management Science*, 37(10):48–68.

Takatsu, S. (1980). Decomposition of satisficing decision problems. *Information Sciences*, 22:139–49.

Takatsu, S. (1981). Latent satisficing decision criterion. *Information Sciences*, 25:145–52.

Taylor, M. (1987). *The Possibility of Cooperation*. Cambridge University Press, Cambridge.

Thaler, R. H. (1986). The psychology and economics conference handbook: Comments on Simon, on Einhorn and Hogarth, and on Tversky and Kahneman. In Hogarth, R. M. and Reder, M. W., editors, *Rational Choice*. University Chicago Press, Chicago.

Thomas, B., Shoham, Y., Schwartz, A., and Kraus, S. (1991). Preliminary thoughts on an agent description language. *International Journal of Intelligent Systems*, 6(5):497–508.

Tribus, M. (1969). *Rational Descriptions, Decisions and Designs*. Pergamon Press, New York.

Tucker, H. G. (1967). *A Graduate Course in Probability*. Academic Press, New York.

Tversky, A. and Kahenman, D. (1986). Rational choice and the framing of decisions. In Hogarth, R. M. and Reder, M. W., editors, *Rational Choice*. University of Chicago Press, Chicago.

von Neumann, J. and Morgenstern, O. (1944). *The Theory of Games and Economic Behavior*. Princeton University Press, Princeton, NJ. (2nd edn., 1947).

Wangermann, J. and Stengel, R. F. (1999). Optimization and coordination of multiagent systems using principled negotiation. *Journal of Guidance, Control, and Dynamics*, 22(1):43–50.

Weibull, J. W. (1995). *Evolutionary Game Theory*. MIT Press, Cambridge, MA.

Weiss, G., editor (1999). *Multiagent Systems*. MIT Press, Cambridge, MA.

Wellman, M. (1993). A market-oriented programming environment and its application to distributed multicommodity flow problems. *Journal of Artificial Intelligence Research*, 1:1–23.

Wellman, M. and Doyle, J. (1991). Preferential semantics for goals. In *Proceedings of AAAI-91*, pages 698–703.

Whitehead, A. N. (1937). *Adventures in Ideas*. Macmillan, London.

Whitehead, A. N. (1948). *An Introduction to Mathematics*. Oxford University Press, London.

Widrow, B. (1971). Adaptive filters. In Kalman, R. and DeClaris, N., editors, *Aspects of Network and System Theory*, pages 563–87. Holt, Rinehart and Winston, New York.

Wiener, N. (1949). *The Extrapolation, Interpolation and Smoothing of Stationary Time Series*. Wiley, New York.

Wierzbicki, A. P. (1981). A mathematical basis for satisficing decision making. In Morse, J. N., editor, *Organizations: Multiple Agents with Multiple Criteria*, pages 465–83. Springer-Verlag, Berlin.

Wigner, E. (1960). The unreasonable effectiveness of mathematics in the natural sciences. *Communications in Pure Applied Mathematics*, 13(1).

Wolf, G. and Shubik, M. (1974). Concepts, theories and techniques: Solution concepts and psychological motivations in prisoner's dilemma games. *Decision Sciences*, 5:153–63.

Wolpert, D. H. and Tumer, K. (2001). Reinforcement learning in distributed domains: An inverse game theoretic approach. In *Proceedings of the 2001 AAAI Symposium: Game Theoretic and Decision Theoretic Agents*, pages 126–33. March 26–28, Stanford, CA. Technical Report SS-01-03.

Wu, S.-H. and Soo, V.-W. (1999). Game theoretic reasoning in multi-agent coordination by negotiation with a trusted third party. In *Proceedings of the Third Annual Conference on Autonomous Agents*, pages 56–61.

Zadeh, L. A. (1958). What is optimal? *IRE Transactions on Information Theory*, 4(1):3.

Zadeh, L. A. (1965). Fuzzy sets. *Information and Control*, 8:353–88.

Zeleny, M. (1982). *Multiple Criteria Decision Making*. McGraw Hill, New York.

Zeng, D. and Sycara, K. (1997). Benefits of learning in negotiation. In *Proceedings of AAAI-97*, pages 36–41.

Zeng, D. and Sycara, K. (1998). Bayesian learning in negotiation. *International Journal of Human Computer Systems*, 48:125–41.

Zilberstein, S. (1998). Satisficing and bounded optimality. In *Proceedings of the 1998 AAAI Symposium*, pages 91–4. March 23–25, Stanford California. Technical Report SS-98-05.

Zlotkin, G. and Rosenschein, J. S. (1990). Negotiation and conflict resolution in non-cooperative domains. In *Proceedings of the Eighth National Conference on Artificial Intelligence*, pages 100–5.

Zlotkin, G. and Rosenschein, J. S. (1991). Cooperation and conflict resolution via negotiation among autonomous agents in noncooperative domains. *IEEE Transactions on Systems, Man, and Cybernetics*, 21(6):1317–24.

Zlotkin, G. and Rosenschein, J. S. (1996a). Compromise in negotiation: Exploiting worth functions over states. *Artificial Intelligence*, 84(1-2):151–76.

Zlotkin, G. and Rosenschein, J. S. (1996b). Mechanism design for automated negotiation and its application to task oriented domains. *Artificial Intelligence*, 86(2):195–244.

Zlotkin, G. and Rosenschein, J. S. (1996c). Mechanisms for automated negotiation in state oriented domains. *Journal of Artificial Intelligence Research*, 5:163–238.

Name index

Subject index